Engineering Design:
The Conceptual Stage

Engineering Design:

The Conceptual Stage

M. J. French
M.A., M.Sc., M.I.Mech.E.
Professor and Head of Department of Engineering
University of Lancaster

Heinemann Educational Books
London

Heinemann Educational Books Ltd

LONDON EDINBURGH MELBOURNE TORONTO
SINGAPORE JOHANNESBURG AUCKLAND
IBADAN HONG KONG NAIROBI NEW DELHI

ISBN 0 435 71650 6

First published 1971

Published by Heinemann Educational Books Ltd
48 Charles Street, London W1X 8AH

Printed in Great Britain at
the Pitman Press, Bath

Preface

THIS BOOK is about the early or conceptual stages of design—more specifically engineering design, but many of the ideas put forward are equally applicable to all kinds of functional design and so will be of interest to industrial designers and others.

Much of the book requires little or no engineering knowledge, and all of it is within the grasp of a second-year engineering student. The demands made on the reader's knowledge are small, but those made upon his intelligence are less so.

It is hoped that reading this book and wrestling with the problems in it will be of great benefit to young engineers, not only those who later become designers but also those who will be involved in creative engineering in other roles, such as research and development, or who will supply specialist functions like stress analysis in support of design. It should also help those who, as managers, will have to make decisions about design policy.

Further, it is felt that the practising designer, to whom most of the ideas will be at least somewhat familiar, will find this book of interest and perhaps of value in strengthening and refining his approach. And if one such, reading it, says here, 'I know a better example than this,' and there, 'But hasn't he realized you can look at such problems so?,' and sits down and writes a better book, that will be so much the more justification for this one.

It is not the purpose of this book to add to the reader's stock of theory or useful facts; its purpose is to cultivate an attitude of mind, a discipline of thought, and a set of approaches to design problems. To this end, many peripheral topics, such as the management of design and the relationship of design to other functions, have been left out. Many books, ostensibly on the same subject, deal almost entirely with these fringe activities, consigning the central issue to a blackbox mysteriously inscribed 'devise solutions,' or something of the kind: this fault at least has been avoided.

While not quite a textbook, this book should make good background reading for engineering students, and engineering teachers will find it contains much matter on which exercises for design courses can be based. There are over seventy problems with sketch solutions and notes, on a very wide range of topics.

To economize in background explanations, the same example has often been used several times to illustrate different points. No apology should be needed for this device, since it reflects the value in design, as in other fields, of looking at the same thing from several points of view.

M.J.F.

Units and Symbols

Units

Mostly SI units are used. In a few cases British units or both kinds are given. For reference, unit abbreviations which may be unfamiliar are listed here.

A ampere
ft foot
in inch
J joule (= newton-metre)
K degrees Kelvin
kg kilogramme
kWh kilowatt hour
lbf pound force
m metre
N newton (= kg m/s^2)
s second
W watt (= joule/s)
Wb weber
p.s.i. pounds force per square inch
p.s.i.a., p.s.i.g. pounds per square inch absolute and gauge
S.H.P. shaft horsepower

For the convenience of those not yet familiar with the newton, 1N $\simeq$ 0.2248 lbf. Pressures and stresses are given mostly in newtons per square millimetre and 1N mm^{-2} $\simeq$ 145.0 p.s.i.

Symbols

Because of the large variety of topics and the brevity of their treatment, notation has been introduced and explained in the text. Customary usage has been followed in appropriate cases.

Questions

A number of questions (Q) are given at the end of each chapter, except 1 and 8, together with outline answers (A) or notes (N) on their solution. The average standard is rather high, and it is not expected that the reader will be able to solve many of them without the help provided. In some cases the question number is followed by a number between 1 and 5 intended to indicate difficulty; e.g. Q.4.5(3) indicates the fifth problem on Chapter 4 and that it is of the third order of difficulty.

The questions are mostly intended as worked examples forming an extension of the text, throwing new light on the subject matter and indicating ways of applying the methods and ideas advanced, so that the reader who wishes to go more deeply into the subject should certainly study them. The questions at the end of Chapter 10 cover matter from throughout the book.

Contents

	PREFACE	v
	UNITS AND SYMBOLS	vii
	QUESTIONS	ix
Chapter 1	**Introduction**	
1.1	Design: conceptual design: schemes	1
1.2	The anatomy of design	1
1.3	The scope and nature of design methods	4
1.4	Insight	5
1.5	Diversification of approach	6
1.6	Reduction of step size	7
1.7	Prompting of inventive steps	8
1.8	Generation of design philosophies or synthesizing principles	8
1.9	Increasing the level of abstraction	8
Chapter 2	**Combinative Ideas**	
2.1	Introduction	10
2.2	Construction of tables of options: functional analysis	11
2.3	Functional analysis: axial flow compressor rotor	12
2.4	Parametric mapping of viable options	14
2.5	Liquid natural gas tanker: alternative orders	19
2.6	Further examples of combinative treatments	23
2.7	Elimination procedures for tables of options	25
2.8	Analysis of interconnected decision areas	28
2.9	Elimination procedures (*continued*)	30
2.10	Synthesis from tables of options: kernel tables	32
2.11	Evolutionary techniques: hybrids	34
2.12	Evolutionary techniques: redistribution of functions	35
2.13	Repêchage	39
2.14	Combinative ideas: general remarks	40
	Questions	40
	Notes	41
Chapter 3	**Optimization**	
3.1	Introduction	44
3.2	Making capital and running costs commensurate	44

3.3 Optimum speed of a tanker 45
3.4 The optimization of the sag/span ratio of a suspension bridge 47
3.5 Optimization with more than one degree of freedom—heat exchanger 49
3.6 Putting a price on heat-exchanger performance 53
3.7 Variation of costs with application 55
3.8 Further aspects of heat-exchanger optimization 55
3.9 An elementary programming problem 56
3.10 Classification of optimization problems and methods of solution 58
3.11 The design of rotating discs: an optimum structure 63
3.12 Hub design 66
3.13 Importance of optimization in design 70
Questions 70
Answers and notes 72

Chapter 4 Insight
4.1 Introduction 74
4.2 Rough calculations 74
4.3 Optimization of compressor shaft diameter 81
4.4 The optimum virtual shaft: a digression 84
4.5 Useful measures and concepts 85
4.6 Bounds and limits 90
4.7 Scale effects 93
4.8 Dimensional analysis and scaling 97
4.9 Proportion 99
4.10 Change of viewpoint 100
Questions 103
Answers 105

Chapter 5 Matching
5.1 Matching: the windlass 108
5.2 An extended example of matching: ship propulsion 108
5.3 Matching within a single machine 112
5.4 Further aspects of ship propulsion 114
5.5 Specific speeds: degrees of freedom 115
5.6 Matching of a spring to its task 117
5.7 Matching in thermodynamic processes 120
5.8 Two old cases of matching 124
5.9 Car handbrake 126
5.10 General remarks on matching problems 130
Questions 130
Answers 134

Chapter 6 Disposition

6.1 Problems of disposition 137
6.2 Dovetail fixing 137
6.3 Splined shaft 138
6.4 Disc brake caliper: breaking logical chains 139
6.5 Alternator rotor 141
6.6 L.N.G. tanker 143
6.7 A digression: ship structures 145
6.8 L.N.G. tanker (*resumed*) 147
6.9 The design of gear teeth 152
6.10 Disposition of allowable stress 155
6.11 Electromagnets 156
Questions 157
Answers 159

Chapter 7 Kinematic and Elastic Design

7.1 Introduction 163
7.2 Spatial degrees of freedom: examples of kinematic design 163
7.3 Elastic design and elastic pairs 171
7.4 Comparison of kinematic and elastic means 177
7.5 Elastic design and structural economy 180
Questions 183
Answers 185

Chapter 8 Tasks, Costs, Assets, and Values

8.1 Introduction 190
8.2 Tasks and values in gearing 191
8.3 Gears for contra-rotating propellers 194
8.4 Large pressure vessels for gas-cooled reactors 196

Chapter 9 Various Principles and Approaches

9.1 Introduction 198
9.2 Avoiding arbitrary decisions: degrees of freedom of choice 198
9.3 Mathematical models 200
9.4 The search for alternatives 203
9.5 Logical chains 204
9.6 Past practice and changed circumstances 206
9.7 Brainstorming 206
Questions 207
Answers 209

Chapter 10 Conclusion and Select Bibliography

10.1 Resumé 210
10.2 Checking and evaluation 211
10.3 Bibliography: general works on design 213

10.4 Bibliography: special aspects of design 213
10.5 Bibliography: case studies 214
10.6 Bibliography: background reading 214
Questions 215
Answers 220

REFERENCES 231

INDEX 233

1
Introduction

1.1 Design: conceptual design: schemes

In this book design is taken to mean all the process of conception, invention, visualization, calculation, marshalling, refinement, and specifying of details which determine the form of an engineering product.

Design generally begins with a need. This need may be met already by existing designs; in such cases the designer hopes he can meet the need better (i.e. generally more cheaply). It ends with a set of drawings and other information to enable the thing designed to be made. Ideally the intervening stages should be of successively increasing precision, of gradual crystallization or hardening. The early stages differ in character somewhat from the later ones, largely because of the greater fluidity of the situation. These early stages, when there are still major decisions to be made, are called *conceptual design* [1]. The products of the conceptual design stages will be called *schemes.*

By a *scheme* is meant an outline solution to a design problem, carried to a point where the means of performing each major function has been fixed, as have the spatial and structural relationships of the principal components. A scheme should be sufficiently worked out in detail for it to be possible to supply approximate costs, weights, and overall dimensions, and the feasibility should have been assured as far as circumstances allow. A scheme should be relatively explicit about special features or components but need not go into much detail over established practice.

1.2 The anatomy of design

Figure 1.1 is a block diagram showing the design process according to the writer. The circles represent stages reached, and the rectangles represent work in progress.

Constructing block diagrams is a fashionable pastime, especially in fields like design where boundaries are imprecise and interactions legion, so that any ten experts will produce ten (or a hundred). They will all be different, and all valid—'there are nine and sixty ways of constructing tribal lays, and every single one of them is right.' They express only truisms, and yet they have a value for all that.

In this particular block diagram things that are important but outside the scope of this book have been omitted—for example, relationships with other

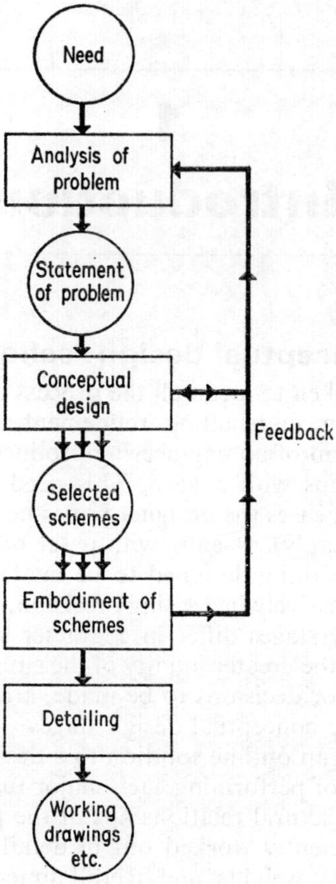

Figure 1.1 Block diagram of design process

activities like research and development, inputs of information and so forth. On the other hand, there is no box labelled 'evaluation' because the writer believes it should be going on continuously in all the rectangles.

Analysis of the problem

This part of the work consists of identifying the need to be satisfied as precisely as is possible or desirable. For such things as consumer goods or widely used items of machinery, it is a problem of locating and defining an adequate 'ecological niche' (section 10.6) and this is the hardest case.

The analysis of the problem is a small but important part of the overall process. The output is a statement of the problem, and this can have three elements:

(1) a statement of the design problem proper,
(2) limitations placed upon the solution, e.g. codes of practice, statutory requirements, customers' standards, date of completion etc.,
(3) the criterion of excellence to be worked to.

Now the ultimate criterion of excellence is always cheapness; all such things as reliability, efficiency and so forth can be reduced to costs, given sufficient information, and always should be as far as possible. Even matters of safety ought to be examined on a basis of cost-effectiveness. The best design is the cheapest, provided that the cost is properly assessed.

However, this ultimate criterion of low cost is often inconveniently remote, and more immediate criteria are sometimes of greater use. For example, in designing an aircraft gas turbine it is useful to know the equivalent capital value of saving a pound in weight so that a limit is set within which the lightest design is the cheapest.

An important point to notice about the analysis of the problem is that there is feedback to it from the design work proper. Few needs are absolute; most are relative to the costs of filling them. The design of a private car, for example, might well develop in such a way as to change its original statement of problem. Features which were excluded as likely to make the price too high might prove cheaper than was expected, and others originally regarded as essential might be so irreconcilable with the developing design philosophy (see section 2.10) as to become unduly expensive.

Conceptual design

This phase of the design process forms the main topic of this book. It takes the statement of the problem and generates broad solutions to it in the form of schemes. It is the phase that makes the greatest demands on the designer, and where there is the most scope for striking improvements. It is the phase where engineering science, practical knowledge, production methods, and commercial aspects need to be brought together, and where the most important decisions are taken.

Embodiment of schemes

In this phase the schemes are worked up in greater detail, and if there is more than one, a final choice between them is made. The end product is usually a set of general arrangement drawings. There is (or should be) a great deal of feedback from this phase to the conceptual design phase, which is why the writer advocates overlapping the two.

Detailing

This is the last phase, in which a very large number of small but essential points remain to be decided. The quality of this work must be good, otherwise delay and expense or even failure will be incurred: it is reasonable to hope that computers will fairly soon effect an improvement here and both reduce the drudgery and improve the standards of this skilled and patient work.

This brief description of the design process will suffice for present purposes. Apart from organizational problems, the difficulty of maintaining the quality of detail draughting, and so forth, there are three central problems in design:

(a) the generation of good schemes (conceptual design),
(b) securing the best embodiment of those schemes (the problem of best embodiment), and
(c) the evaluation of alternatives.

This book is primarily about the first problem, but inevitably becomes concerned with the others. Also, many of the approaches suggested for generating schemes are directly applicable to the embodiment stage as well, since the design of a part may be tackled much as if it were a complete design problem.

1.3 The scope and nature of design methods

This book is an attempt to show what design methods can do to help the designer, principally in the conceptual stage: it will no doubt appear very amateurish in future years when the subject has been thoroughly developed.

First, let us be clear what design methods cannot do and probably never will be able to do. They cannot replace the gifts of the talented designer, nor provide step by step instructions for the production of brilliant designs. What they may perhaps do is to improve the quality and speed of the able designer's work, and increase the size and range of the tasks he can tackle. They may also speed the development of the young designer and, perhaps most important of all, improve the co-operation with essential specialists inside and outside the design office.

The ways in which design methods may help can be roughly classified as follows:

(1) by increasing insight into problems, and the speed of acquiring insight,
(2) by diversifying the approach to problems,
(3) by reducing the size of mental step required in the design process,
(4) by prompting inventive steps, and reducing the chances of overlooking them, and
(5) by generating design philosophies (synthesizing principles, design rationales) for the particular problem in question.

Other benefits may be expected from the keeping of better records of the progress of design work, from wider-spread and more accurate knowledge of the reasons for past decisions, even from less ambiguous and more rapid communication between designers.

Not all design methods will be useful when applied to any given problem: an analogy may help to explain this. Consider a boy trying to prove a rider in Euclidean geometry. He knows a number of theorems, and he has to find those which will help him in his particular problem. The problem itself provides some clues—there is a circle with a tangent and a chord, so he applies the theorem about the angle between a tangent and a chord. This probably does not solve his problem, it merely presents him with a changed situation that may suggest a further step. The kind of talent the boy requires in applying theorems is rather like that which the designer needs in applying methods. As with the designer, there are generally many solutions to the boy's problem (which is to find a *proof*, not a result) some elegant, others clumsy but adequate. Like the designer, he will feel at times he is making progress, and sometimes the feeling will be justified and sometimes it will not.

Design problems, like other tough creative problems, are solved by sheer hard mental work. Methods help the designer to keep himself working, to tide over usefully those periods when inspiration will not come, or to break out of circular arguments. They help him to grasp the eel-like tail of a solution as it flashes through the corner of the imagination and pin it out on the drawing board, when, alas, it often turns out to be unworkable. Even then, however, method will often mean it is more quickly seen to be useless.

The methods that are discussed in the rest of this book, which are sometimes called ideas, approaches, techniques, or aids, according to their degree of formalization or the width of their applicability, have certain things in common. They involve a removal of the problem to a higher level of abstraction; an attempt to clear away the purely circumstantial and to look for a general element which will widen the field of known means in which a solution can be sought; an attempt to substitute measures for merely qualitative words like advantage and disadvantage, dear and cheap; in short, the sort of bricks with which reason has always built.

1.4 Insight

It is a common experience to find that an idea, worked on long and hard, has some basic flaw which at a later stage is instantly recognized, because insight

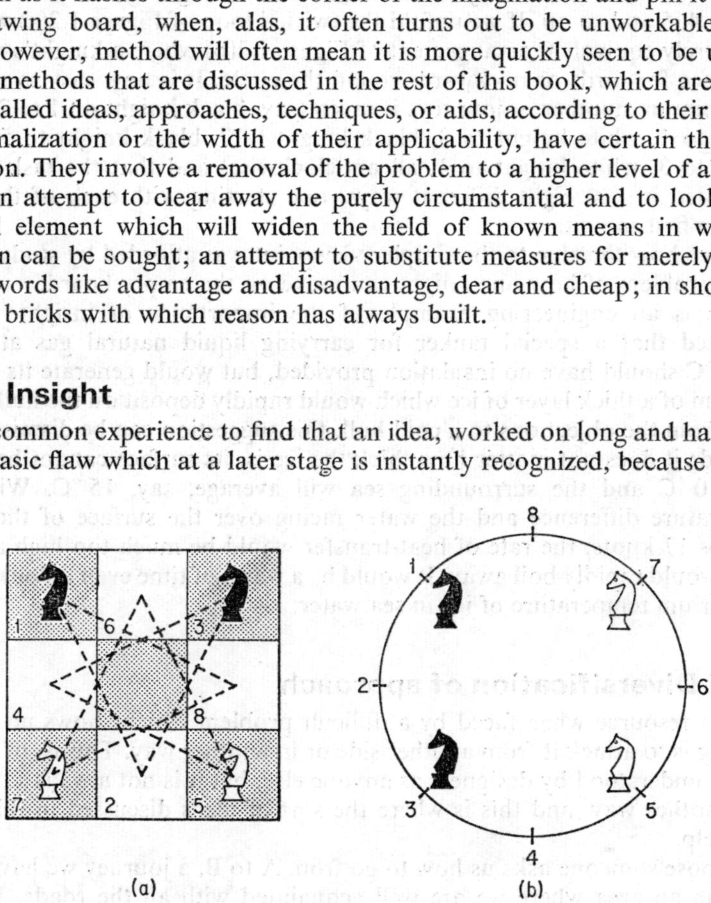

Figure 1.2 Knights problem

into the problem has improved. It is therefore wise for the designer tackling a new problem to spend some of his time improving his insight into it as rapidly as possible. This will not be done by careful studies along one set of lines, but by many rapid exercises, looking at different aspects in different ways, a sort of familiarization procedure.

The ancient problem illustrated in Figure 1.2(a) may serve to illustrate the value of insight and how it may be gained by looking at the problem in a different way. The black knights and the white knights have to be exchanged, subject to the ordinary rules of chess for moving knights, by means of the minimum number of journeys, where a journey is any number of consecutive moves by the same knight. Now if we remove three knights and move the remaining one continually, never going back to the square just left, it traces out a closed path. If we start with the black knight on the square numbered 1 and move him to 2, his only move then, apart from going back, is to 3, and then to 4, 5 and so on. If we unfold the twisted loop of Figure 1.2(a) we have the entirely equivalent arrangement of Figure 1.2(b) where a knight's move is one space forwards or backwards round the circle. It is now easy to see that the minimum number of journeys is seven (say, black knight on 1 to 2, white knight to 1, white knight to 8, black knight to 7, black knight to 5, white knight to 3, white knight to 1, all anticlockwise) and that the task can be performed in just eight different ways, one starting with each of the eight possible first moves.

Notice how the clue to the simpler viewpoint was afforded by studying the trivial problem of how one knight can move on the empty board.

Here is an engineering example of the importance of insight. It was suggested that a special tanker for carrying liquid natural gas at about −160 °C should have no insulation provided, but would generate its own in the form of a thick layer of ice which would rapidly deposit on the hull. Quite apart from the objections to a cold hull, this suggestion can be dismissed out of hand: it does not matter how thick the ice is, its surface cannot be above about 0 °C and the surrounding sea will average, say, 15 °C. With this temperature difference and the water racing over the surface of the ice at perhaps 17 knots, the rate of heat-transfer would be much too high and the cargo would rapidly boil away. It would be a waste of time even to look up the equilibrium temperature of ice in sea water.

1.5 Diversification of approach

A great resource when faced by a difficult problem which shows no sign of yielding is to attack it from another side or in another way. This is, of course, as well understood by designers as anyone else, but it is not always so easy to find another way, and this is where the sort of ideas discussed in this book may help.

Suppose someone asks us how to go from A to B, a journey we have never made in an area where we are well acquainted with all the roads. We can answer much more readily with a map before us, though it does not add anything essential to our knowledge. We can at once see all the alternative routes which otherwise would take time to think out, and all the considerations like level crossings or congested areas which influence the choice between them and might otherwise be forgotten. So design methods can prompt ways of tackling problems which otherwise might not be hit upon or, more probably, hit upon so late that the work they would have saved has already been wasted.

Very often, the designer has difficulty in bringing his mind to bear profitably on the problem in hand: release from this unhappy state is often brought by a happy thought, a minor inspiration, which causes him to readjust slightly his point of view, and progress is resumed. One of the ideas in this book might well have brought that happy thought at less cost in time and brain-cudgelling. It has often been suggested that such seemingly unprofitable hours are a necessary precursor of the leap forward (e.g. reference 2) but the writer does not believe this is true of the small steps that have to be made every day.

Viewed from a single aspect, a design problem often shows too great a variety of possibilities: walk around it, and certain imperatives are borne in on us, some constraints which restrict the areas in which a solution may lie. Design is a catch-as-catch-can business—we must seize a problem wherever there is something to take hold of. For example, in a particular heat-exchanger problem, where the range of forms which might have proved cheapest appeared at first sight very large, the consideration of what the baffles had to be like and the resulting difficulties of assembly pruned the possibilities much more satisfactorily than any other aspect.

1.6 Reduction of step size

Some great ideas appear so obvious it is difficult to understand why they were not thought of earlier. Sometimes a contributory reason for this is that hindsight has, as it were, a series of unacknowledged stepping stones whereby to cross the stream that inspiration has leapt. The great idea can be explained in a series of steps, none of which in itself appears remarkable, and once such an explanation has been given, it is difficult ever again to grasp how wide the gap was. The moral is that those of us who are not good long-jumpers should look for handy stones when we wish to cross streams.

Among the small steps into which the arguments leading to good and profitable design innovations may be broken, we may expect some to be rather special to the problem in question, but others to be of a more general kind, the sort that might be made as part of a routine approach. Various steps of this sort occur in the examples discussed in this book.

One elementary but useful way of reducing step size may help to give the general idea. If a difficult function has to be performed, we can make some progress by listing the means we can think of which meet some of the requirements and then tackling the separate smaller problems of adapting each of these to meet the full requirements: with any luck we shall be able to solve one or more of these smaller problems, perhaps by introducing an *auxiliary function*, or by recognizing that the original function may be divided into two (see section 2.12). To give a trivial example, we might need to support a large load in a region at a high temperature, and this might be done by using a member suitable only for a lower temperature and surrounding it by a refractory shroud with cooling air flowing through it.

1.7 Prompting of inventive steps

The claim that method may prompt inventive steps will seem rash, if not ridiculous, to some. But arguments which can be built on the lines put forward here will often reduce to a marching logic which leads inexorably to a minor but unmistakeable invention: all this, it seems to the writer, can happen without any very great mental leaps.

It is difficult to illustrate this. The field in which the writer has been most aware of this possibility is that of thermodynamic process design, where the problems are relatively tidy and precise, the available means strictly limited and the rules of the game well understood and immutable. In this specially favourable case it is reasonable to imagine a computer, as it were, nudging the designer and saying 'Look at this, look at these particular values, shouldn't there be an improvement possible here?' Indeed, the writer has had patented ideas produced in just this way, except that he had to go through the figures himself to find the prompting values.

The Linde-Fränkl process for the separation of atmospheric oxygen and nitrogen is an example of marching logic leading to a major inventive step in the thermodynamic process field [3].

1.8 Generation of design philosophies or synthesizing principles

Mention has been made in the last section of marching logic which leads the designer inexorably onwards. However, a vague but helpful indication of direction is all that can be expected in most cases. A number of examples are given later, together with some account of how such design philosophies can grow out of systematic treatments and formal approaches.

1.9 Increasing the level of abstraction

Two simple precepts the designer should always bear in mind are 'quantify whenever possible, even if only roughly' and 'increase the level of abstraction at which problems are treated until it ceases to be profitable'.

Both these precepts are extensively illustrated later. The second, however, warrants some enlargement at this point because it is an important characteristic of almost all the ideas put forward.

All mental disciplines must rest heavily on the use of abstract concepts, which are essential to intellectual processes. Moreover, progress in understanding is often associated with increasing levels of abstraction. But abstraction may take different forms, and it is important in design that the right forms should be selected, especially as specialist writings in the engineering field are mostly analytical. As a result there is a grave shortage of good creative abstractions; those that exist are not widely known, and accepted terminologies are virtually non-existent.

As an example, it seems unlikely that many readers are familiar with the term 'corner power' (see section 4.6), and yet this is a most helpful concept in many fields of design. Sensible heat (enthalpy rise associated with temperature rise) has been banished from the thermodynamic vocabulary, perhaps rightly, but no substitute has appeared. The analyst may be able to avoid the use of such a term without much circumlocution, but for the designer the loss is serious. Many thermodynamic processes suffer from what the writer calls 'sensible heat cycling', where a large 'water equivalent' (another abstraction, which is useful in design and is falling into disfavour) is subjected to large temperature cycles. Sensible heat cycling nearly always involves large losses as, for example, in Newcomen's engine before Watt improved it (section 2.12). Stirling-cycle engines are a rare example where sensible heat cycling is nearly reversible. It is a great impediment to the discussion of the design of a thermodynamic process not to be able to say something like: 'but if we do that we shall increase the sensible heat cycling', and be understood.

Another useful concept is 'holding-together power', the product of the tensile stress and the volume in a tensile structural material. This, like corner power, is an example of a measure of a task to be performed or the ability to perform a task. If a pressure vessel has a volume V and has to stand an internal gauge pressure p then its holding-together power is $3pV$ (see section 8.4).

More abstract does not always mean more general. If we want to design an elastic beam, the highly abstract but very specialized view of the beam as two flanges and a web, the flanges taking all the moment and the web all the shear, is immeasurably more useful than the very general theory of elasticity. The key to the superiority of the cruder concept here is its greater abstraction (it has only three areas and a depth, at most, defined at any section) and its *purpose-related* nature. The beam has a *resistant function* (Torroja), which is to resist a certain pattern of shear and bending moment over a certain length. We provide separately for the two parts of its function, a web, to resist shear, and two suitably spaced flanges to sustain the bending moment.

The kinds of abstraction specially useful in design are generally those associated with purposes, with tasks or the ability to perform tasks, with limitations or with losses.

2
Combinative Ideas

2.1 Introduction

A design problem might be tackled on the following lines: take all the possible elements that might enter into a solution, combine them in all possible ways, and then select the best combination.

This ideal combinative process has three important characteristics: the synthesis aspect, such as it is, is reduced to a mechanical process of systematic

ADVANTAGE	COMBINATION Ff	Fr	Rf	Rr
Less flexure to be accommodated in shafts driving wheels		✓		✓
Better road holding due to forward C of G	✓	✓		
Better road holding due to front wheel drive	✓		✓	
No propeller shaft, clear floor space	✓			✓

Figure 2.1 Table of options, motor-car configuration

combination, and the difficulties are reduced to those of analysis; it appears impossible to overlook good solutions; lastly, it is impracticable.

The chief reason for the impracticability of this comprehensive approach is the overwhelming number of possibilities inherent in almost all design situations. Put another way, the number of degrees of freedom of choice is much too large for all variations to be studied on their individual merits. There is some analogy with certain games. In noughts and crosses all the possible plays may be traced to their conclusion and the best play in any position determined. It is a game which can be treated exhaustively, or is 'exhaustible' for short. The writer is not expert enough to be sure if draughts is exhaustible—it is near the borderline perhaps—and chess, to date, is far from exhausted. Practically speaking, design problems are inexhaustible.

Nevertheless, much advantage may be drawn from the central idea of a combinative procedure, used sparingly in any of a number of different ways. As a very simple example, consider the design of a private car. We can put the engine at the front, F, or the rear, R, and we can drive the front wheels f, or the rear wheels r, giving four possible arrangements

Ff Fr Rf Rr

The principal advantages of these different arrangements are set out in a table in Figure 2.1. It can be seen that every scheme is superior to every other scheme in one or more respects, except Rf which has only one of the three advantages possessed by Ff and no other advantage. Rf may thus be regarded as wholly inferior to Ff and can be rejected, without attempting even to put numerical values to the various conflicting advantages. This is in accordance with existing practice, in which there are plentiful examples of the other three configurations but none of Rf. To proceed farther than this one easy discard, very detailed studies are needed.

Before discussing the uses of combinative procedures in conceptual design we shall examine the resources they offer.

2.2 Construction of tables of options: functional analysis

The essence of all combinative methods is some sort of table of options, of which Figure 2.1 is a very small example. Since the number of different combinations is often too large for each to be represented by a separate pigeonhole, the table is usually condensed by displaying only the separate options.

Various kinds of option generally exist, and they may be viewed in different ways, so that it is possible to construct tables on several bases, of which the chief are:

(1) alternative means of performing essential functions,
(2) alternative solutions to various smaller design problems or *sub-problems* [46] arising within the main one,
(3) a classification (virtually, a taxonomy) derived from a study of known solutions of the same or allied problems,
(4) alternative spatial arrangements, configurations or orders (Figure 2.1 is an example),
(5) mixtures of some of the first four.

The choice of basis must depend on the nature of the problem and the preference of the designer.

Basis (1) is suitable for all new problems where no extensive background exists, but it can also help to throw new light on old problems.

Basis (2) is well adapted to cases with an extensive background, where the difficulties are well understood and the dominant sub-problems have emerged clearly. It is economical of effort in that it concentrates attention on the vital areas. Care must be taken, however, that the scope of enquiry is not limited to the point where radical alternatives are overlooked.

Basis (3) is of fundamental interest, in that the systematic tool is generated out of the results of non-systematic design. It can be used in very difficult cases in which it would be hard to make headway by any other procedure, and the table it produces is automatically well suited to the task in hand.

Basis (4) is most likely to be of use in parts of problems, or for studying a single aspect of a large problem.

2.3 Functional analysis: axial flow compressor rotor

An important tool for constructing tables of options is a *functional analysis*, which sets out the essential functions that the design is required to perform.

Consider the problem of the mechanical design of the rotor for an axial flow compressor, all the blade dimensions being already determined; such rotors are shown in Figures 2.4–2.8.

The principal functions to be filled by the rotor are:

(1) Providing an attachment for the rotor blades,
(2) Transmitting the drive to the rotor blades,
(3) Resisting centrifugal loads due to the rotor blades,
(4) Fairing the inner wall of the gas duct,
(5) Providing adequate transverse stiffness.

The even-numbered functions here will be largely ignored. Although essential, (2) is a trivial problem to provide, and (4), though it has an important bearing, would complicate the example too much if properly taken into account.

The loads involved in (3) are very large relative to the size of the rotor, and make both (1) and (3) exacting problems. A large transverse stiffness (5) is needed to avoid shaft vibration or whirling troubles; we will adopt here the basically unsound but practically justified procedure of ensuring a minimum lowest transverse natural frequency of vibration of the rotor in rigid bearings as a precaution against whirling.

In forming a table of options for an axial flow compressor rotor design, functions (2) and (4) will be omitted. We proceed to find all the likely means we can of performing the remaining functions. It is unlikely that any new means will be hit upon—all the options that go into the table will probably come from what the Patent Office calls 'known means'.

For example, the attachment of the blades to the rotor, which for simplicity are assumed to be demountable, can be done by the well-known bulb, dovetail or fir-tree fixings, or by pins or rivets, or by circumferential grooves, as illustrated in Figure 2.2.

It is worth making a small digression at this point to study the way in which a classification can be made of blade attachments; it brings out several points of interest.

One broad division is into those solutions that require extra parts and those which do not. The former require extra load transfers from part to part, and so, on a task-cost basis (section 8.1), they must carry an inherent weight penalty, which may, however, be offset by higher working stresses or which may be justified by advantages such as reduced vibratory stress levels. Those which do not have extra parts have sliding engagement (another kind, having convergent engagement, is rare and unlikely, e.g. taper pipe threads and carpet sweeper handles), and so their mating surfaces must be composed of lines whose geometric properties are identical at all points in their length. Such lines must have constant curvature and torsion (rate of rotation of the plane of curvature about the tangent to the line), i.e. they must be helixes. A

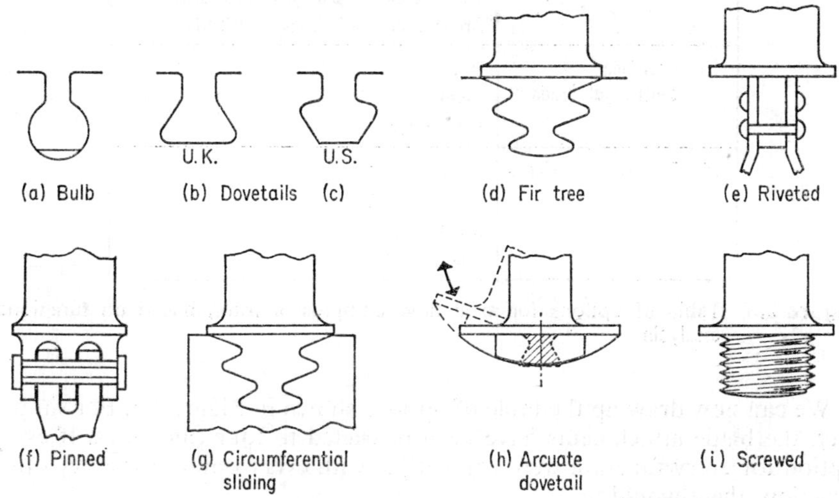

Figure 2.2 Forms of blade root fixing for axial flow compressors

helix with zero torsion is a circle, and a helix with zero curvature is a straight line, giving the three kinds of sliding engagement associated with the three lower kinematic pairs; the screw pair, the turning pair, and the sliding pair. This is exhaustive—there are no other kinds.

It is useful to note in listing alternative blade attachments such characteristics as the effect they have on possible rotor constructions—straight-line sliding roots are usually impracticable with solid rotors—or the amount of dismantling necessary to change blades.

Function (3), resisting centrifugal loads, is treated exhaustively in section 3.11. Here we shall consider three options, the disc, the ring or drum, and the solid rotor, the disc being about twice as economical in material as a ring or drum of thin radial extent.

Function (5) is similar to that of a bridge, and any form of bridge structure that is independent of anchoring and abutment thrust requirements would serve, but in practice only a beam or shaft is reasonable. One important choice that does remain is whether the shaft is a continuous piece, i.e. a separate shaft, or a stack of pieces butted end to end and prestressed by a central bolt or tendon or a ring of bolts, an arrangement which will be called a 'virtual' shaft. Figure 2.6 shows such a rotor construction. Finally, a drum or a solid rotor with blades inserted directly in it will serve also as a shaft.

FUNCTION	MEANS
(1) Blade attachment	Straight sliding (Figure 2.2 a, b, c, d) With additional parts (Figure 2.2 e, f) Circumferential sliding (Figure 2.2 g) Arcuate dovetail (Figure 2.2 h)
(3) Resisting centrifugal loads	Disc Drum Solid rotor
(5) Transverse stiffness	Separate shaft Virtual shaft Drum Solid rotor

Figure 2.3 Table of options for axial flow compressor rotor, based on functiona analysis

We can now draw up the table of options shown in Figure 2.3. For simplicity, the blade attachments have been restricted to four categories. If every option for every function were compatible with every option for every other function, there would be

$$4 \times 3 \times 4 = 48$$

possible combinations. However, functions (3) and (5) lumped together give only four options, solid rotor, drum, discs and separate shaft, and discs and virtual shaft, reducing the total number of combinations to 16.

In the general case we can get no further. All the options are likely to be viable in certain circumstances: what is required next is some delimitation according to circumstances, and to achieve this we shall use two parameters, one measuring the difficulty of functions (1) and (3), and the other measuring the difficulty of function (5).

2.4 Parametric mapping of viable options

If r is a typical radius, say, a radius to the root of a rotor blade about half-way through the compressor, and ω is the angular velocity of the rotor, $\omega^2 r$ may be taken as a typical centripetal acceleration and $\rho\omega^2 r^2$ as a measure of centrifugal stress level (see section 3.11). If f is a typical allowable stress, perhaps as in Figure 3.10, the ratio

$$P_{13} = \frac{\rho\omega^2 r^2}{2f} \tag{2.1}$$

is a measure of the relative difficulty of providing adequate strength against centrifugal forces; the reason for the 2 will appear in section 3.11.

The difficulty of providing adequate transverse stiffness will depend on ρ, a typical density, and E, the Young's modulus of the shaft, virtual shaft or drum material. They will appear in a suitable parameter in the form ρ/E or some power of ρ/E. The form preferred here is $(\rho/E)^{1/2}$, which is the inverse of the velocity of sound in the shaft material if it has density ρ. The complete form of the parameter is

$$P_5 = k\frac{L^2}{r^2}u\sqrt{\left(\frac{\rho}{E}\right)} \tag{2.2}$$

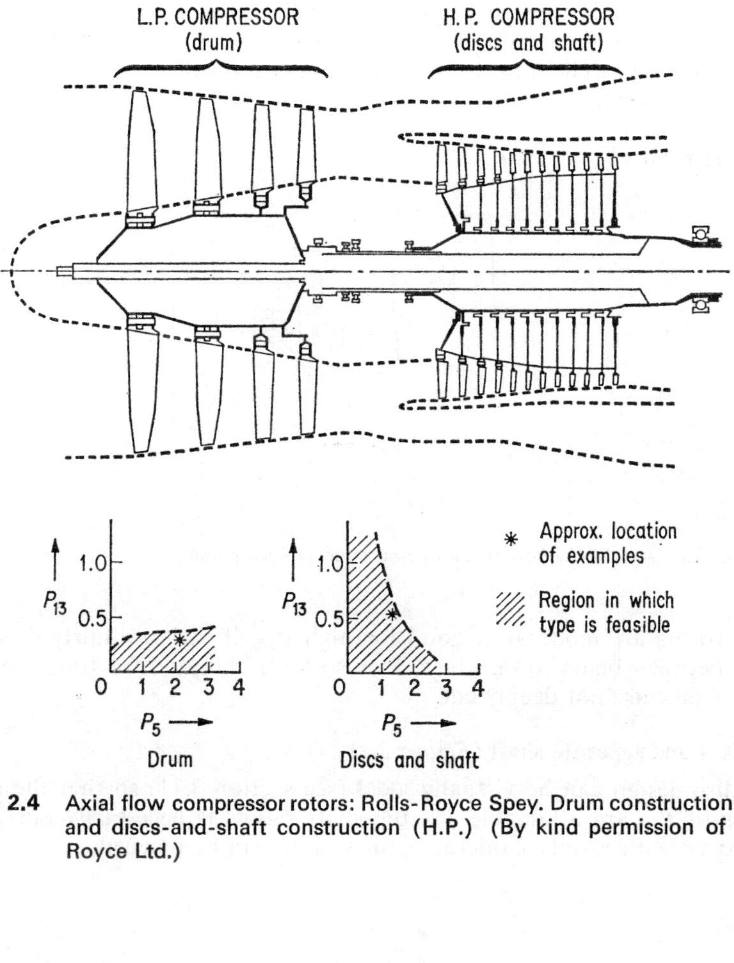

Figure 2.4 Axial flow compressor rotors: Rolls-Royce Spey. Drum construction (L.P.) and discs-and-shaft construction (H.P.) (By kind permission of Rolls-Royce Ltd.)

where u is the blade speed, L is the length of the rotor between bearing centres, and k is the ratio of the required lowest natural frequency to the running speed. P_5, like P_{13}, is a dimensionless parameter; it is the product of a factor k, the ratio L^2/r^2, which is a measure of the slenderness of the rotor and clearly makes the stiffness requirement harder to meet as it increases, and the ratio of the blade speed to the speed of sound in the material. The whole thing is a measure of the difficulty of the function (5) requirement, an incomplete measure since it does not allow for certain other variables like hub/tip ratio, but good enough.

We can now plot against our two parameters P_{13} and P_5 the areas in which the various options are likely to be suitable, as has been done in Figures 2.4–2.7. The term 'solid rotor' is here restricted to mean rotors machined from the solid but with a solidity of about one half or more.

Drum (*Figure 2.4*)

The drum solution is impossible above $P_{13} = 0.5$, for then a ring of radius r is just able to sustain its own weight without the stress exceeding f (see section 3.11). Drum solutions will scarcely be viable above $P_{13} = 0.25$. On the other hand, they are excellent for transverse stiffness and will go to as high values of P_5 as are possible.

Solid rotor (*Figure 2.5*)

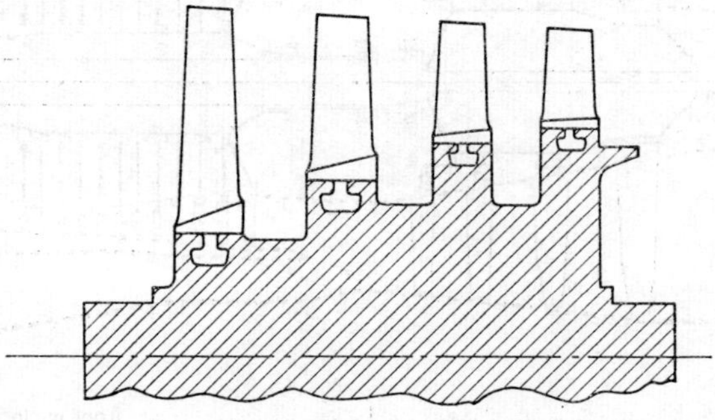

Figure 2.5 Axial flow compressor rotor, solid (hypothetical)

Solid rotors are moderately good for high P_{13}, if they are fairly deeply cut away between blade rows. They can also be fairly good for transverse stiffness, if they are not deeply cut.

Discs and separate shaft (*Figure 2.4*)

The disc design can be virtually ideal (see section 3.11) so that the highest values of P_{13} are attainable. As the shaft radius is necessarily fairly small compared with r, only moderate values of P_5 can be reached.

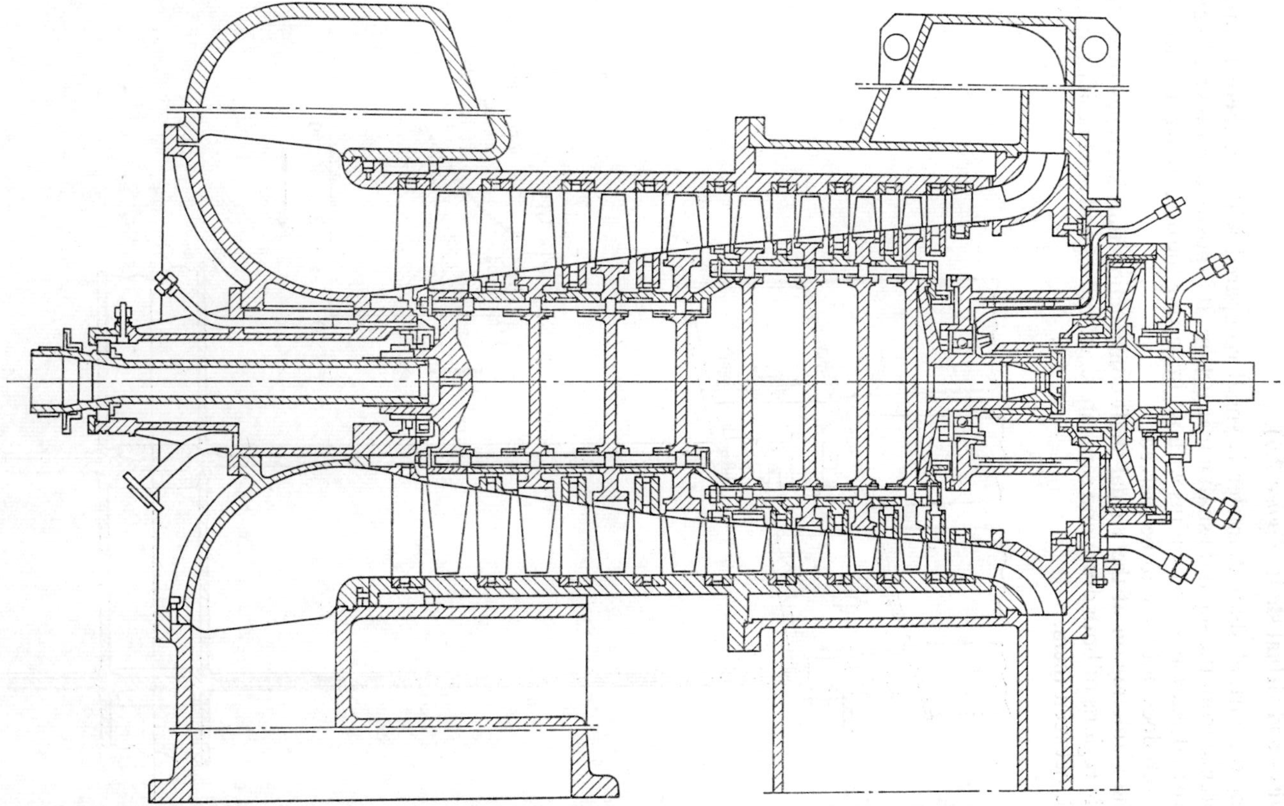

Figure 2.6 Axial flow compressor rotor, discs-and-virtual-shaft construction. (Based, with kind permission, on a figure from *Thermal Engineering*, a translation of the Russian journal *Teploenergetika*; 1966, **13**, 5)

Discs and virtual shaft (*Figure 2.6*)

When the virtual shaft diameter is small compared with $2r$, there is little difference from the separate shaft construction. It is possible, however, to put the virtual shaft at a large radius without impairing the disc strength, since the disc web does not have to be pierced.

The viable region of the virtual-shaft construction thus stretches over the viable regions of both drum and separate shaft and a good area besides in the angle between them.

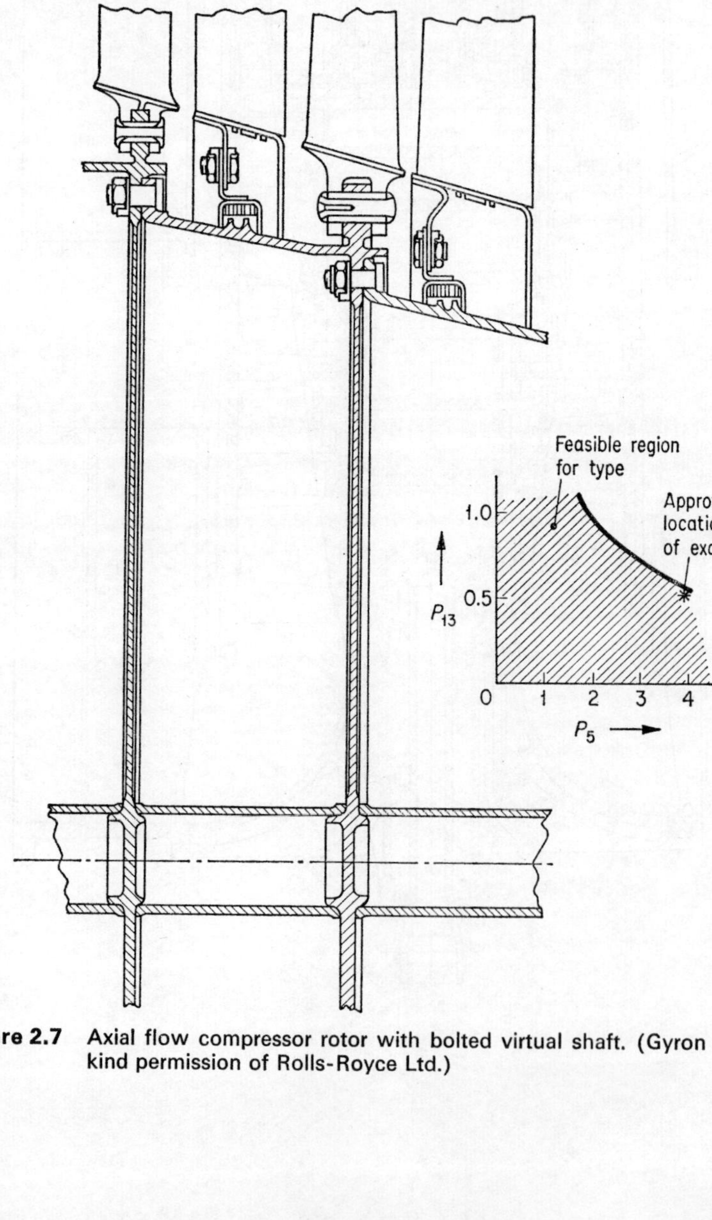

Figure 2.7 Axial flow compressor rotor with bolted virtual shaft. (Gyron Junior: by kind permission of Rolls-Royce Ltd.)

We have now some idea of the regions of the 'parameter-space' P_{13}, P_5, in which different constructions are viable. We could go much further, e.g. to examine the effect of size, measured perhaps by the radius r, that makes constructions like that of Figure 2.7 possible for very large compressors. We could consider the effect of the alternative forms on cost of manufacture and weight, which would help to show which were more suitable for aircraft and land use. But enough has been done to show the general picture.

One last point is this. The classification used here becomes a little woolly when designs for a high P_5 and a high P_{13} (say, around 3 and 0.5) are concerned. The only suitable constructions may be regarded as discs-and-virtual shaft with a very large diameter virtual shaft, or else as *hybrids of discs and drum*, with just sufficient drum to provide the necessary transverse stiffness (Figure 2.8). The drum part may be able to carry a little of the centrifugal load, but most of it is carried by the disc webs. The design process here may be regarded as drawing the independent optimum structures to carry out functions (3) and (5), and then marrying them together in a manufacturable whole with as little departure from ideality as may be. The first part of this procedure can be entirely formalized ('exhausted') (see sections 3.11 and 4.4). It is to be expected that more and more engineering products will lend themselves to such approaches as the technological targets we set ourselves become more exacting.

2.5 Liquid natural gas tanker: alternative orders

This example concerns the special tankers used to transport liquefied natural gas (L.N.G.) at substantially atmospheric pressure and about −161°C (112K). The tanks are thermally insulated and the heat which leaks in boils off the natural gas at an acceptably low rate, the vapour being used as part of the fuel.

A trivial functional analysis of the tank section of such a tanker shows the following principal requirements:

Containment of L.N.G.	C (i.e. preventing leakage)
Thermal insulation of L.N.G.	I
Structural resistance to pressure of L.N.G.	S
Normal hull function	H

These four functions can be visualized as residing in up to four layers from inside to out. We can use a convenient code for the possible configurations, e.g.

C (IS) H

means that a layer providing containment C is innermost, followed by a single layer combining the functions of insulation I and structure S, and so on. There are in all 75 such arrangements.

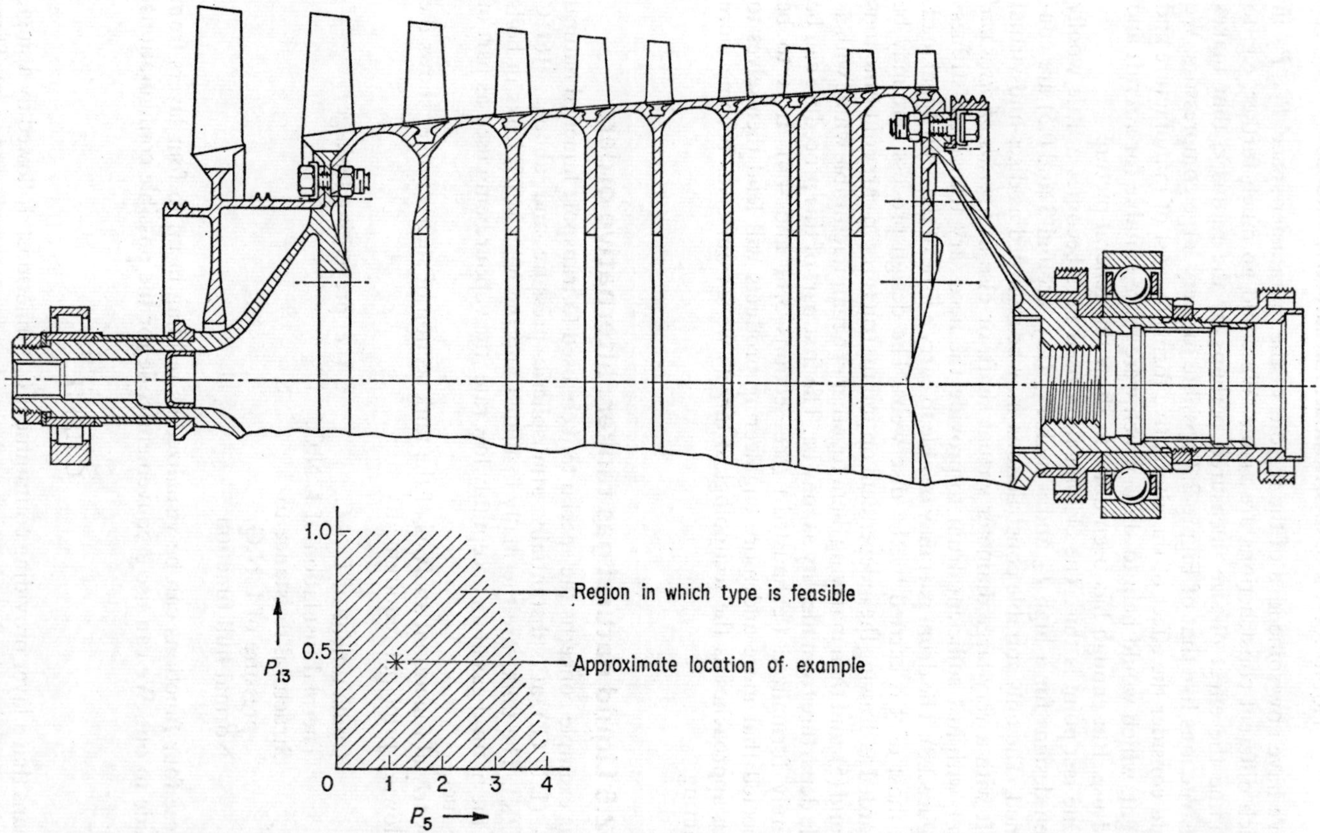

Figure 2.8 Axial flow compressor rotor, one-piece discs-and-virtual-shaft. (Gnome: by kind permission of Rolls-Royce Ltd.)

Codes of this sort are useful, especially for concise and unambiguous communication between designers and for recording the results of design work. Figure 2.9 shows a possible construction corresponding to the code (CS)IH.

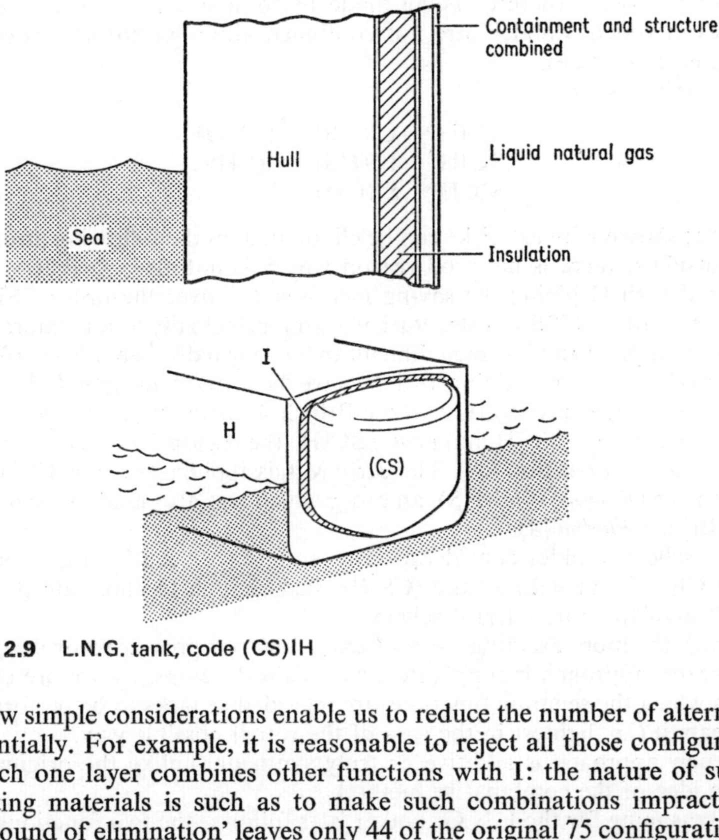

Figure 2.9 L.N.G. tank, code (CS)IH

A few simple considerations enable us to reduce the number of alternatives substantially. For example, it is reasonable to reject all those configurations in which one layer combines other functions with I: the nature of suitable insulating materials is such as to make such combinations impracticable. This 'round of elimination' leaves only 44 of the original 75 configurations. If the I layer is outside the layer supplying the H function, the hull will be at roughly −160 °C, which has among its other overwhelming disadvantages this: the hull will have to be made in a material free from brittle fracture down to that temperature, and the cheapest of such materials is much dearer than conventional ship-building steels, apart from increased fabrication costs. It is also difficult to imagine either C or S lying outside H, so that all orders which do not have H as one function of the outer layer can be rejected: this leaves 13 survivors.

Another elimination may be made on these lines: the two structural functions, S and H, involve resisting internal and external pressures respectively. If the two are combined in one structure, the structural tasks (see section 8.1) will partly cancel each other and the material required for both functions will be less than for H alone. There is such a large saving here that no scheme which does not take advantage of it where possible can be regarded as viable. The only reason this *combination of functions* may not be possible in some cases is that I separates S and H, so that one is in a cold region and the other at ambient temperature, and a problem of differential expansion prevents the two structures being made to co-operate. We therefore reject schemes in which S and H are not combined although not separated by I, leaving eight schemes.

The survivors are

CSIH*	CI(HS)	(CS)IH
SCIH	IC(HS)	I(CHS)
SICH*	SI(CH)*	

The three shown with asterisks can be eliminated on the following grounds: if C is outside I, there is no good reason why S is not also, so that S can be combined with H giving the saving mentioned above; the order CSIH can only be economical if S is a structure working principally as a membrane (i.e. *not* a plate or grid) and it is then difficult to see why it does not also perform C.

Four of the five remaining orders have been used in actual ships. The French *Jules Verne* is (CS) IH and the British *Methane Princess* and *Methane Progress* are partly (CS)IH and partly SCIH: the reason for this hybridization is explained later (section 6.8). The early Mississippi barges were I(CHS) and the French *Pythagore* is CI(HS), an order which was also used experimentally in the British *Findon* [4].

Some schemes under consideration recently have been hybrids which are mostly CI(HS) but with a little (CS)IH, just enough to eliminate the worst difficulties of the pure CI(HS) scheme.

As with the more exacting cases of axial flow compressor rotor design, the combinative approach is supplanted, as insight develops, by a more abstract view in which the separate functions are regarded as tasks to be performed in the cheapest (i.e. lightest in the case of the rotor) possible way.

The new approach is additive or truly synthetic, unlike the original subtractive idea of the combinative method.

The reasoning for the L.N.G. tanker is as follows: the two functions S and H taken together are an absolutely smaller structural task than H taken separately, since the internal pressure of the L.N.G. will help resist the external pressure of the sea. Numerical values in joules, say, can be put on these tasks. It is therefore cheapest, other things being equal, to combine S and H, particularly since if S is cold (as in the (CS)IH schemes) expensive materials must be used to avoid brittle fracture. We therefore *synthesize* the two preferred orders CI(HS) and I(CHS). The latter fails to meet practical safety requirements, leaving CI(HS): the difficulty with this scheme is that if the lining performing function C is made to fit the insulated hold in the first place, it will contract away from the walls when the very cold L.N.G. is pumped in. If, say, a simple stainless-steel lining were used, the pressure of the rising L.N.G. would thrust it back against the walls and burst it. The CI(HS) scheme can only be realized if some way of overcoming this differential expansion problem can be found. Briefly, the following are the possibilities:

(a) Use a material for C which can be constrained to stay with the walls even when cold, without excessive stresses being set up (e.g. invar, which exhibits very little thermal contraction: see Question 4.3).
(b) Devise a form of corrugated sheet which can be stretched equally in all directions (see reference 4).
(c) Use some 'active' element between the lining and the hold walls that will follow up the thermal contraction of the lining and continue to support it even in its 'shrunk' position.
(d) Use a hybrid, mostly CI(SH) but a little (5–15 per cent) (CS)IH, as suggested above, just as the exacting compressor rotor case is met by a disc solution with just the right admixture of drum.

While (c) is not a very promising runner, it leads to a further search for alternatives.

2.6 Further examples of combinative treatments

Examples have been given of combinative treatments based on the analysis of function and on configurations or orders. Figure 2.10 shows various configurations of reduction gear for turbine-driven tankers, together with comparative costs and weights [5]. This is a good example of a study which has been carried through for all the alternatives in a fair amount of detail.

Figure 2.11 illustrates the Systematic Design Method of Rodenacker, as applied to the design of a machine used in the quality control of foil production. It punches small discs from a roll of foil according to a predetermined pattern, weighs them automatically and transmits the data for processing. A part of the method is a combinative study in which both functions and configurations are treated and Figure 2.11 shows the selected optimum means for the foil sampling machine. Figure 2.12 shows a trial assembly of these optimum aspects into a first rough scheme. Rodenacker's Design Institute at the Munich Technische Hochschule has done much valuable work, for the plastics industry chiefly, and is now making its name in other fields. His design method has been developed to a high level and its usefulness thoroughly proved in practical applications (section 10.3).

The best example known to the writer of a classification forming the basis of a table of options is that used by Wankel, the inventor of the N.S.U. Wankel engine which will be known to most readers. There have been numerous inventions of 'chamber mechanisms', i.e. mechanisms in which a chamber formed between two or more parts cyclically increases and decreases in volume. Subject to certain practical limitations, such mechanisms can be used as bases of expansion engines, i.e. steam engines, internal-combustion engines, air motors etc., in place of the familiar piston and cylinder arrangement. Wankel formed a systematic classification for all such inventions using only rotating components, with about 800 different combinations of options or pigeon holes.

	GEAR ARRANGEMENT	Reduction	HP	IP	LP	HP	IP	LP	Height	Width	Length	Weight	Price
				K			RPM			INCH		RELATIVE	
1	HP	1	125	–	125	6000	–	3600	260	299	142	1.40	1.20
	LP	2	95	–	87								
2	HP	1	125	–	125	6000	–	3600	240	280	114	1.18	1.24
		2	95	–	78								
		1	150	–	150				228	268	106	1.00	1.05
	LP	2	125	–	102								
3	HP	1	300	–	300	6000	–	3600	213	327	130	1.21	1.12
		2	95	–	87								
		1	300	–	300				201	311	126	1.00	1.03
	LP	2	125	–	109								
4		1	300	–	–	6000	–	3600	189	272	142	1.00	1.00
		2	300	–	300								
		3	95	–	95								
	HP	1	300	–	–				173	256	142	0.90	0.89
		2	300	–	300								
		3	125	–	125								
		1	300	–	–	8000	–	3600	189	272	142	1.00	0.97
		2	300	–	300								
	LP	3	95	–	95								
		1	300	–	–				173	256	142	0.90	0.85
		2	300	–	300								
		3	125	–	125								
5		1	300	–	–	6000	–	3600	201	240	138	0.93	1.06
		2	125	–	125								
		3	95	–	95								
	HP	1	300	–	–				193	232	130	0.77	0.92
		2	150	–	150								
		3	125	–	125								
		1	300	–	–	8000	–	3600	201	240	138	0.93	1.06
	LP	2	125	–	125								
		3	95	–	95								
		1	300	–	–				193	232	130	0.77	0.92
		2	150	–	150								
		3	125	–	125								
6	HP	1	300	300	300	11000	7000	3600	157	244	142	0.80	1.00
		2	300	300	300								
	IP	3	95	95	95								
		1	300	300	300				146	240	142	0.66	0.91
	LP	2	300	300	300								
		3	121	121	125								
7		1	300	–	–	6000	–	3600	146	189	177	0.55	0.85
		2	95	–	95								
		3		300									
	HP	1	300	–	–				134	177	177	0.50	0.82
		2	125	–	125								
		3		300									
		1	300	–	–	8000	–	3600	146	189	177	0.55	0.85
		2	95	–	95								
	LP	3		300									
		1	300	–	–				134	177	177	0.50	0.82
		2	125	–	125								
		3		300									
8	HP	1	300	300	–	11000	7000	3600	157	189	177	0.62	0.88
		2	73	67	95								
	IP	3		300									
		1	300	300	–				146	177	177	0.56	0.86
		2	99	88	125								
	LP	3		300									
9	HP	1	300	–	300	6000	–	3600	150	220	256	0.90	1.01
		2_o	95	–	82								
	LP	2_i	95	–	87								

KEY

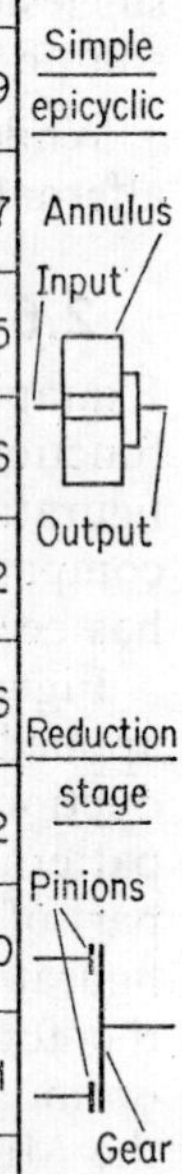

Now in such a classification, some of the practical considerations, like the compression ratio that can be achieved, difficulties of sealing and providing valves, inherent balance or lack of balance and so forth, are associated with the pigeon holes rather than the particular designs [6]. Wankel devised numerous new chamber mechanisms, concentrating on those that occupied the most promising pigeon holes, and one of these was the basis of the successful design which is now being widely adopted.

2.7 Elimination procedures for tables of options

A table of options constructed in one of the ways suggested in section 2.2 may be regarded as a large number of schemes in embryo. In some cases, such as the tanker gearing shown in Figure 2.10, it may be practicable to study all the feasible combinations reasonably thoroughly, but in the more typical case, like the L.N.G. tanker, only a few variations can be studied in depth, and most of them must be eliminated by methods of a more cursory nature. Here it is convenient to consider the tables based on functional analysis separately from those based on order.

Occasionally it may be possible to decide on the best means of performing each function separately, and by combining these options obtain the best combination, i.e. the optimum solution is also the optimum in respect of each function considered independently. More often, however, the best means selected in this way will be incompatible, or more expensive to combine, so that the best overall solution will incorporate some other alternatives.

In the foil sampling machine (Figures 2.11 and 2.12), it proved possible to optimize each means separately. This is often the case with plant that carries out of a series of processes, where the number of kinds of interaction between one function and the ones that precede and follow it are limited. We shall describe such design situations as of weak or limited interaction.

As a simple example of strong and complex interaction, consider the various advantages listed in the table in Figure 2.1. The first three are

---

Figure 2.10 Combinations of marine gearing with two or three turbine shafts: results of a comparative design study.
The diagrams in the left-hand column indicate the configuration under study, e.g. the third diagram down shows the turbine divided into two sections, H.P. and L.P., each driving into a simple epicyclic. The output shaft of each epicyclic drives a pinion meshing with a single large gear on the propeller shaft. The last diagram shows a version for contra-rotating propellers, in which drives are taken from both the output shafts and the annulus gears of the epicyclics (see section 8.3). Studies have been made of the effects of three degrees of freedom—division of the turbine into two or three sections (see sections 5.3–4) different materials (shown by the different K's) and different H.P. turbine speeds—besides the different gear arrangements. The factors K are a measure of tooth stress, i.e. they are proportional to the constant C in equation 4.13.
(By kind permission of Dr. I. Jung and the Society of Naval Architects and Marine Engineers)

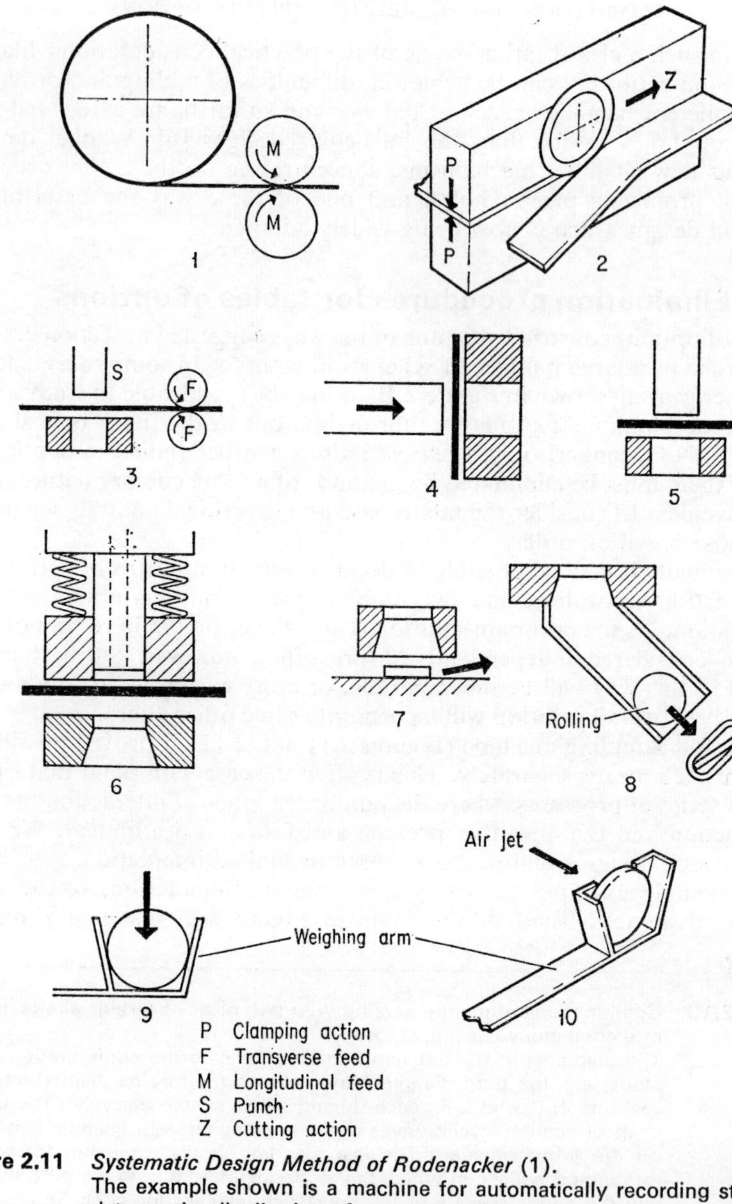

Figure 2.11 *Systematic Design Method of Rodenacker* (1).
The example shown is a machine for automatically recording statistical data on the distribution of mass per unit area in a sample roll of foil, for quality-control purposes. The data are obtained by punching out of the foil and individually weighing a pattern of small discs. The diagrams illustrate the separately optimized means of performing the functions 1–10, where 1 is feeding lengthways, 2 is cutting in widthwise strips, 3 is feeding the strips sideways to the punch which punches out the discs (functions 4, 5, 6) etc.

associated with particular options r, F and f respectively. But the fourth advantage, no propeller shaft and a clear floor space, is not associated with a particular option—indeed, the two combinations, Ff and Rr, which share it have no common option. In the language of the statistician, this particular advantage is a *first-order interaction*. In the same way, a *second-order interaction* would be an advantage, disadvantage or other consequence associated with three options taken together. It is convenient to use the term 'interaction' both in

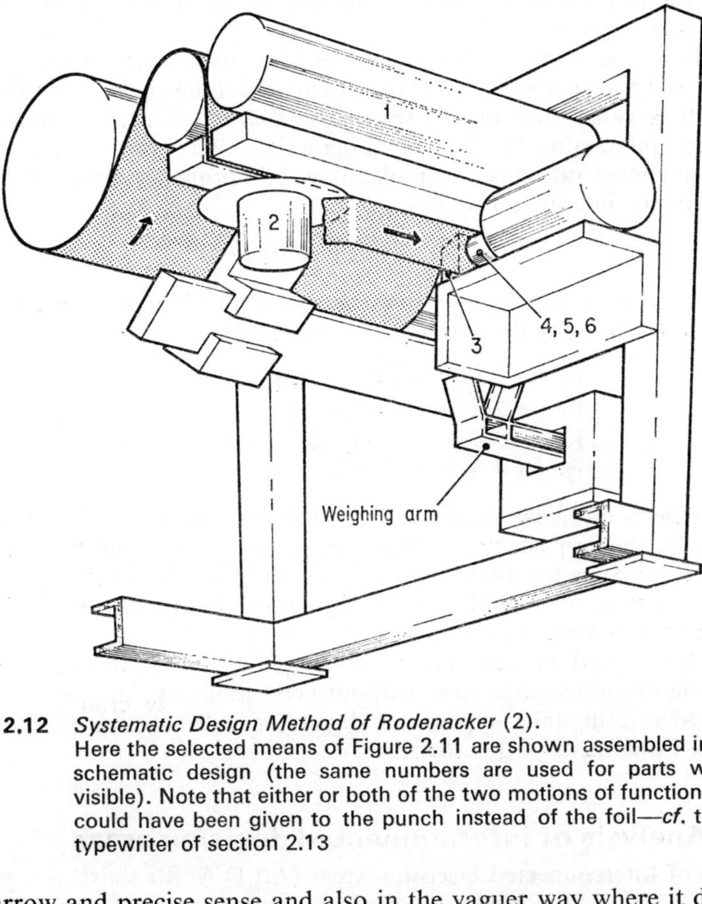

Figure 2.12 *Systematic Design Method of Rodenacker* (2).
Here the selected means of Figure 2.11 are shown assembled into a first schematic design (the same numbers are used for parts which are visible). Note that either or both of the two motions of functions 1 and 3 could have been given to the punch instead of the foil—*cf.* the I.B.M. typewriter of section 2.13

this narrow and precise sense and also in the vaguer way where it describes only the general order of interconnectedness of problems.

One way of finding the best combination is to put *values* on each option and on each interaction and find the total values for all the possible combinations. Indeed, by putting a very large negative value on the interaction of two options which are incompatible, we can reduce the problem to one of finding the highest total value, regardless of whether combinations are possible or not.

Thus, in the case of the motor-car configuration problem, we can consider the values of the two alternative locations of the engine, F and R. The only aspect allowed for in this simple treatment is the effect on road-holding, so that we might put F = 2, R = 0, say. The 'zero position' is of no consequence, so we have taken R as zero quite arbitrarily.

With regard to the drive location, the front-wheel drive is advantageous for road-holding but involves the driving shaft flexing through the large angle needed for steering. Suppose for the moment the disadvantage outweighs the advantage and put f = −1, r = 0.

Finally, we need to consider the interactions. In this very simple problem, there are only two choices to be made, where to put the engine and the drive—what will be called here *two degrees of freedom of choice*: consequently the first-order interaction is the only interaction. We assume that the only important consideration is the disadvantage of a propeller shaft, rated at −2, so that for the interaction we put

$$\overline{\mathrm{Ff}} = \overline{\mathrm{Rr}} = 0,\ \overline{\mathrm{Fr}} = \overline{\mathrm{Rf}} = -2$$

where the bar is used to distinguish the *interaction* $\overline{\mathrm{Ff}}$ from the *combination* Ff.

The total values for the combinations are then:

$$\begin{aligned} \mathrm{Ff} &= \mathrm{F} + \mathrm{f} + \overline{\mathrm{Ff}} = 2 + (-1) + 0 = 1 \\ \mathrm{Fr} &= 2 + 0 + (-2) = 0 \\ \mathrm{Rf} &= 0 + (-1) + (-2) = -3 \\ \mathrm{Rr} &= 0 + 0 + 0 = 0 \end{aligned}$$

This reflects the real position for small cars today crudely but fairly realistically. In the past, the difficulty and consequent expense in providing for large flexure in a front-wheel drive favoured Fr and Rr relatively to Ff, but the introduction of reliable cheap high-angle joints to the engineering repertoire has changed the situation.

What has been done here with a very simple problem can clearly be done in much more complicated cases without great difficulty: the process is also well suited to a digital computer. A digression will be made to examine a formalized treatment of this sort.

2.8 Analysis of interconnected decision areas

Analysis of Interconnected Decision Areas (A.I.D.A. for short) is a generalized formal technique for solving a kind of problem common in various fields, including design. It was originated by Harary, Jessop, Luckman, and Stringer at the Institute for Operational Research and has been developed there by several workers including Morgan [7].

It deals with the situation in which a number of strongly interacting decisions have to be made, where certain choices are incompatible and the overall optimum set of choices cannot be found by making each decision on its own merits, i.e. exactly the problem considered in the last section.

A *decision area* exists wherever there is a choice of any kind, of the means of performing a function, of form, material, working fluid etc.; a decision area is what we have called a degree of freedom of choice. In each decision area there are two or more *options*.

A *strategy graph* is a plane network in which each *decision area* is represented by a circle or block. Thus in Figure 2.13, the decision areas are the six

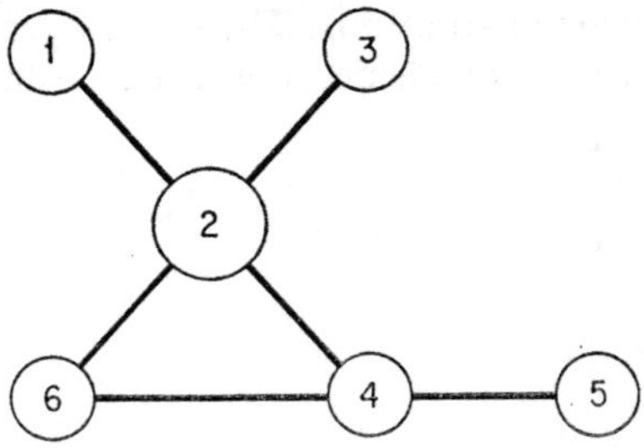

Figure 2.13 A.I.D.A. strategy graph

circles labelled 1 to 6. The line joining decision areas 1 and 2 shows that at least one option in area 1 is incompatible (or compatible only at a significant price) with at least one option in area 2, and so on for all the other pairs of decision areas linked by lines. The layout of the decision areas in the strategy graph should be such as to make the network as simple and readily appreciated as possible: for example, in Figure 2.13 the important 'loop' 246 is clearer than it would be in a straight-line layout with the decision areas in numerical order.

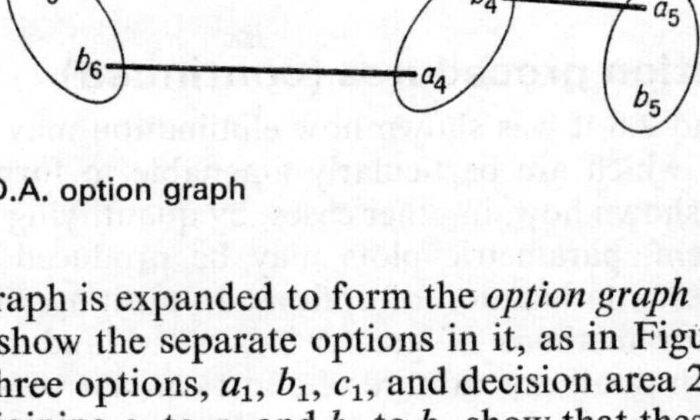

Figure 2.14 A.I.D.A. option graph

The strategy graph is expanded to form the *option graph* by enlarging each decision area to show the separate options in it, as in Figure 2.14. Decision area 1 contains three options, a_1, b_1, c_1, and decision area 2 has two options, a_2, b_2. The lines joining c_1 to a_2 and b_1 to b_2 show that these pairs of options are either wholly incompatible or else that their combination will involve expense or disadvantage not associated with the options taken singly. The option graph helps the situation to be readily grasped and assessed. For example, in Figure 2.14, a little study will show that if the lines represent absolute incompatibility no solution involving b_2 exists (if we choose b_2 in 2, we must choose a_4 in 4 and b_6 in 6: but a_4 and b_6 are incompatible) and all solutions must involve a_2. Since c_1 and b_3 are incompatible with a_2, these options also cannot occur in solutions. The option graph can be reduced to the form of Figure 2.15, in which decision areas 2 and 3, having only one

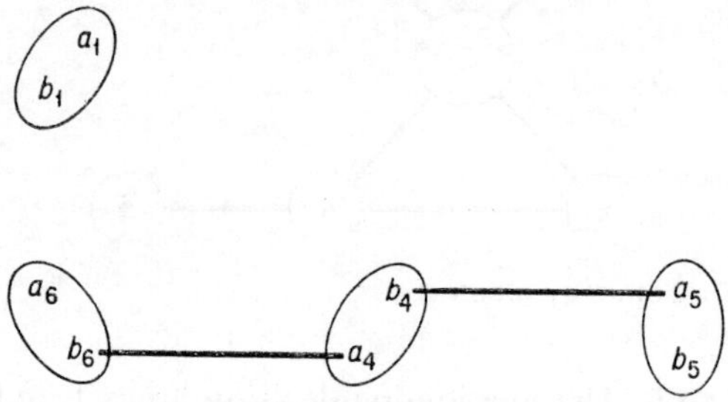

Figure 2.15 Option graph of Figure 2.14 in reduced form

option each left, are no longer needed, and decision area 1 is now quite independent, though it has lost option c_1. It is readily seen there are only eight compatible combinations (a_4, a_5, a_6; a_4, b_5, a_6; b_4, b_5, a_6 and b_4, b_5, b_6, each with either a_1 or b_1).

If the lines do not represent absolute incompatibilities they may represent expensive or unfavourable first-order interactions, and quantities representing costs or values may be assigned to both options and interactions, as in the case of the motor car configuration in section 2.7. The A.I.D.A. method includes computer programs for summing such costs and printing out the solutions or combinations in order of increasing cost or decreasing value. It is also possible to sum more than one kind of cost or value, which is useful if it is desired to postpone making the kinds commensurate.

2.9 Elimination procedures (continued)

In sections 2.7 and 2.8 it was shown how elimination may be performed on tables of options which are particularly amenable to formal treatment. In section 2.3 it was shown how, in other cases, by quantifying certain aspects of the design problem, parametric plots may be produced that help in the selection of the best combination of options. In most applications, however, there is a need for methods of greater flexibility (and hence usually less rigour), which with small expenditure of effort can narrow down the field of enquiry without much risk of missing a good solution.

Such methods must be largely *ad hoc*, devised piecemeal to suit the task in hand, and resting largely on engineering judgement and common sense. Nevertheless, a number of ideas are given in the following notes: in most problems several of these will be of assistance.

(*a*) Grouping of lines of the table of options

Each choice between options corresponds to a line in a table of options written in the form of Figure 2.3, which has three lines corresponding to functions 1, 3 and 5, respectively. It is useful in more elaborate cases to order these lines vertically, so that they fall into groups with most interactions confined within groups and as few and weak interactions as possible between members of different groups. This is the same idea as that of the manipulation of the strategy graph in section 2.8. The mathematical problem involved has been treated by Alexander [8], but a good approach to the ideal restructuring of the table can be made by trial-and-error: the natural structure of the problem is often a great help here, as the groups tend to be related to functional or spatial groups in the real device or system.

(*b*) Amalgamation or condensation of lines

If two lines in a table consist of m and n options, respectively, they can be amalgamated or condensed into a line consisting of the mn combinations they form. This is no improvement unless, as often happens, a substantial number of these mn combinations are impracticable or meaningless. In the axial flow compressor rotor example, the second and third lines of the table in Figure 2.3 have three and four options, but they amalgamate into a line with only four options.

If there are sufficient incompatibilities in the table, whole groups of lines may be amalgamated, e.g. in the option graph of Figure 2.15 the five lines or decision areas numbered 2 to 6 have effectively been amalgamated into a single line with four options (a_4, a_5, a_6; a_4, b_5, a_6; b_4, b_5, a_6; b_4, b_5, b_6) (see section 2.8).

A very satisfactory way for elimination to develop is for several groups to reduce in this way to single lines with only a few options each. To achieve this it is necessary to reject combinations and recognize incompatibilities not only of an absolute kind, but also where quick rough calculations and engineering judgement indicate *prima facie* economic disadvantages too large for there to be much chance of their disappearing on more careful appraisal. One example of this approach was the rejection of orders with S and H adjacent but not combined in the L.N.G. tanker study (section 2.5). An important safeguard in such cases is that the designer should return and reconsider his verdict later, when his insight into the problem has improved.

(*c*) Elimination of single options

Sometimes a short study will suffice to show that one option is inherently inferior to another, whatever the combination in which it is used, and then a slight reduction in complexity can be made by rejecting the inferior option. In the study of disc-like structures in section 3.11, it is shown that spokes or a ring are inferior to a disc as regards economical use of material in resisting large radial loads, by a factor of at least two. In the case of the ring, this is offset by the fact that this structure, extended axially to become a drum, is very good for resisting transverse bending. Consequently a ring or drum design may be viable where the radial strength requirement is not very high and the transverse stiffness needed is large, as was seen in section 2.4. However, the spoked design would appear to have no compensating advantage and might well be rejected, even without considering the severe load-diffusion problems at the hub.

A good practice in such cases is to try and imagine some circumstances in which the option about to be discarded might have some advantage. For example, the spoked wheel might have some virtue for a compressor of very low solidity, say, a very light machine for compressing a gas at a fraction of a pound per square inch, were such a thing ever to be needed (to be convinced of this, study the spoked bicycle wheel and try to find a viable alternative). Another application is to a design using a high-strength filament such as carbon, which is most economically used under uniaxial stressing. Here, no doubt, the outward appearance would be that of a disc, but the true nature might well be spoked.

This is about as much as can usefully be said about elimination without bringing in elaborate notations that are more trouble than they are worth, or giving extended examples. Every user will soon develop his own approach, but a general suggestion may be of use here; it is most economical of effort to be brutal and sweeping to begin with, to arrive quickly at a small group of preferred basic schemes, and later to revise the elimination process in a charitable and open frame of mind, to decide whether some of the eliminated ideas are not worthy of reinstatement; this latter process the writer calls 'repêchage', and it will be referred to again (section 2.13).

However elimination is tackled, two characteristic difficult comparisons arise. The first is between very like solutions and is therefore generally not of the utmost importance: *differential studies* (see section 4.3) are often a help here. The second is between very different solutions, and in the absence of fundamental methods of comparison fairly detailed and expensive design studies may be necessary to resolve the matter. Indeed, if the sum eventually to be spent on hardware is small enough, it will be cheapest to leave the question unresolved and 'plump' for one or the other (see section 10.2).

Where the problems of straight elimination are overwhelming, the techniques of the next section can be used.

2.10 Synthesis from tables of options: kernel tables

We have already seen, in the examples of the axial flow compressor rotor and the L.N.G. tanker, how fuller insight into a problem, treated first by combinative methods, leads to a principle for synthesizing a good design. This state of affairs is to be expected when the technical problems are especially difficult; e.g. the synthesizing principle only becomes apparent and of value when the compressor rotor requires great radial strength combined with high transverse stiffness. It is commonly based upon some quantifiable and readily costed task to be performed by the design, as in the L.N.G. tanker where the structural tasks provide the starting point.

Even when no natural principle of synthesis emerges, we can tackle the problem of selecting preferred solutions from a table of options, not by rejecting the bad, but by building up only the good. In the extreme case, we take some particular line in the table of options and one particular option from that line. We then choose another option from a second line, on the basis of it being the best to combine with the first or *pivotal option.* We then take a third line, and the conditional best option from this line, the conditions being our previous choices from the first and second line, and so on. This process will be called synthesis from a pivotal option. Clearly, its success depends on the choice of the first line and the order in which successive lines are taken; it needs much judgement and is difficult to apply 'from cold', before insight into the particular problem has been developed.

An alternative procedure that combines analysis and synthesis is to construct a small table of options out of a handful of the most important lines in the original table. This small or kernel table is then reduced to a few preferred incomplete combinations by the sort of elimination process already discussed. We now proceed by synthesis as with the pivotal option, but taking one of these preferred incomplete combinations from the kernel table as a pivotal set of options. This is done with each of the preferred incomplete combinations from the kernel table, giving a set of preferred complete combinations. Thus, in the car example of section 2.1 we could regard Figure 2.1 as a kernel table, in which Ff and Rr might be our preferred incomplete combinations. We might then work through a larger table including such choices as form of engine, method of cooling, orientation of crankshaft, types of suspension and so on, so as to end with two schemes, one based on Ff and one on Rr. It is likely, however, that some important choices will be found where the decision on Ff, say, will not make selection any easier, two options, a and b say, seeming to have equal merits; we then increase the number of preferred schemes, pursuing the process with Ffa, Ffb, and Rr.

These synthesizing approaches, where a few good combinations are built up rather than vast numbers of bad combinations being eliminated, seem much more natural and appropriate to the designer. Perhaps the best way is to start with a table of options, to work a little with it by elimination, rejecting altogether inferior options, reordering lines and amalgamating lines, and then to turn to synthesis, based on either a pivotal option or, more probably, a very small kernel table of two or three lines. (The largest problem of this kind that the writer has tackled needed a five-line kernel table, but only five preferred incomplete combinations came out of it). With luck and good management there will emerge a handful of preferred complete combinations, in two or three 'families', with large differences between families and only small differences within them.

It is to be hoped, also, that each family of preferred complete combinations will have its own *design philosophy* or *rationale* or *synthesizing principle*, which the designer or designers will by now understand fairly well. If so, this is the appropriate stage to introduce considerations that are foreign to the original table of options and which would have overburdened it had they been put in at the start. These matters are discussed in the next section.

2.11 Evolutionary techniques: hybrids

The combinative techniques here described are termed 'evolutionary' because they produce new schemes from combinative treatments that do not explicitly contain them. For example, some promising solutions to the L.N.G. tanker problem are based on hybrids, mostly CI(SH) but partly (CS)IH (see section 2.5). Such hybrids are not allowed for in the original permutative scheme, but arise naturally from a conscious synthesizing principle or design philosophy, which in turn grows out of deeper understanding of the problem and the more abstract approach that this understanding makes possible. The processes that produce these hybrids are much more economical than the 'blind' combinative methods since they produce only promising ideas. On the other hand, they are not systematic and must be developed specially for each new problem.

To explain this by an example, let us take the synthesizing principle behind the hybrid CI(SH)–(CS)IH tank. The ideal, already explained in section 2.5, is a thin-walled tank, C, which merely contains the L.N.G. and transmits the pressure load via the insulation I against which it rests to the hull (SH)—this would be a purebred CI(SH) design. The trouble is that the space occupied by the tank or lining C is governed in size by the hull, which remains warm and does not contract when the tank is filled with liquid L.N.G. at about −160°C. If we are to keep a pure CI(SH) design, the lining C must not contract at this temperature, either because it has a very low coefficient of expansion and is stretched the rest, or because it is cunningly corrugated. But suppose the tank does contract but is also deformed in some way; since its superficial area is reduced, it cannot still fit the hold everywhere, but it might fit it *nearly* everywhere. Where it does not fit, it will have to provide structural strength of its own to withstand the cargo pressure; i.e. it will have to be (CS)IH or possibly SCIH (it turns out that we may need 3 to 30 per cent of self-supporting wall, according to materials, type of design, and the size and complexity of tank).

If the tank is basically a polyhedron, the obvious type of deformation is a bulging of the flat faces. Separation takes place at the edges, and they must be made self-supporting or at any rate supported otherwise than by the insulation. Also the 'bulging stresses' must somehow be kept within acceptable limits.

In taking this problem so far, two fairly abstract ideas have been used. The first was the idea of costs associated with quantifiable structural *tasks*, which led us to a strong *a priori* preference for solutions with S and H, the two structural functions, combined in one 'layer'. The second was the hybridizing idea outlined above. Both lead to particular arrangements being adopted—they are synthesizing principles—but the second is narrower and more powerful. By an evolutionary process a design philosophy has grown out of the original combinative treatment.

Note that a common starting point for an evolutionary step, often of the hybridizing sort, is an attempt to produce a scheme to combine the advantages of two other schemes; alternatively, the object may be to avoid both the different weaknesses of its two parents, or to eliminate a single disadvantage from an otherwise promising scheme.

2.12 Evolutionary techniques: redistribution of functions

Improvements in schemes can often be effected by redistributing functions among parts, for example, by using two separate parts to perform two functions that were previously performed by one part. A historic case is Watt's improvement of Newcomen's steam engine. Newcomen used the cylinder of his engine as a condenser also, spraying cold water into it at the end of each stroke. Unfortunately, this cold water not only condensed the steam but also cooled the cylinder walls, so that the first steam admitted on the next stroke was wasted in heating the cylinder again. By using a separate spray condenser alongside the cylinder, Watt reduced the fuel consumption per unit of work by about two-thirds [9].

As an example of the reverse process, of taking two parts each with its own function and combining them into a single part to perform both functions, consider the widespread substitution of a 'unit construction' car body for the earlier system of chassis and separate 'coachwork'.

There are also cases where one function is divided among several parts, as in multi-engined aircraft, and cases where one part with its own proper function helps with the function of another part, as when the structural concrete of a nuclear reactor for power generation helps also to provide the shielding.

Most cases of redistribution of function are largely matters of common sense, and detailed discussion of them would be trivial; the important point for the designer is that he should not overlook the possibilities this technique affords. However, it is probably worth examining a few of the less obvious aspects and such general indications (in the medical sense) as are available.

(1) Distribution of two functions, one to each of two parts, when one part might have done both

This is perhaps the possibility that is most often overlooked when it should not be, sometimes because the two separate functions involved are not recognized as such. In nuclear reactors with reinforced-concrete pressure vessels, the structural function resides in the prestressed steel reinforcing tendons and their concrete matrix, but the containment or sealing function is filled by a thin steel shell (see Figure 2.16). This is a cheaper solution than

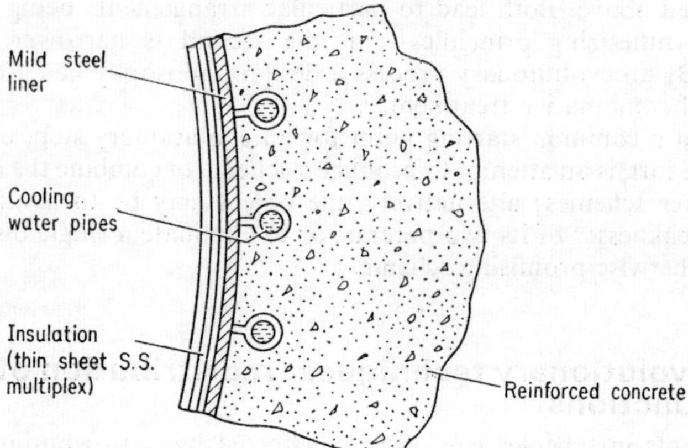

Figure 2.16 Wall of pressure vessel, gas-cooled reactor

combining the two functions in a thick steel shell, as was done earlier, so much so as to make it reasonable to put the whole gas circuit inside one pressure vessel and sweep away many knotty problems of duct design, differential expansion, and so forth.

A close parallel is the Philips hermetic piston seal shown in Figure 2.17. The seal is a 'rolling diaphragm', which works like a sock being turned inside out. Such diaphragms can be made to stand some tens of atmospheres pressure difference by reinforcing with fibres, but the number of reversals which can then be sustained is not sufficient for many applications.

Philips use an unreinforced polymer that can only stand a small pressure difference at which, however, it will survive hundreds of millions of reversals. The space outside the diaphragm is filled with oil at nearly the same pressure as that of the gas inside that is being kept hermetically sealed, so that the pressure difference over the diaphragm is small. The major pressure drop occurs over the outer oil seal, which, of course, leaks oil, but this, unlike gas, can easily be pumped back. This seal is used on Philips Stirling cycle and reversed Stirling cycle engines and refrigerators, where the working fluid is a small quantity of high-pressure hydrogen or helium in the cylinders which should not be allowed to escape, even slowly.

Notice the way in which the cylinder is stepped so that the volume of the oil space in the seal does not vary; this enables the elasticity of the diaphragm to nearly balance the pressures on its two sides.

The Philips C70 reversed Stirling cycle refrigerator is recommended for study as an example of refined and elegant design, combined with a proper regard to cost, as shown by the use of crankcase, crankshaft and connecting rods from an existing design of air compressor [10].

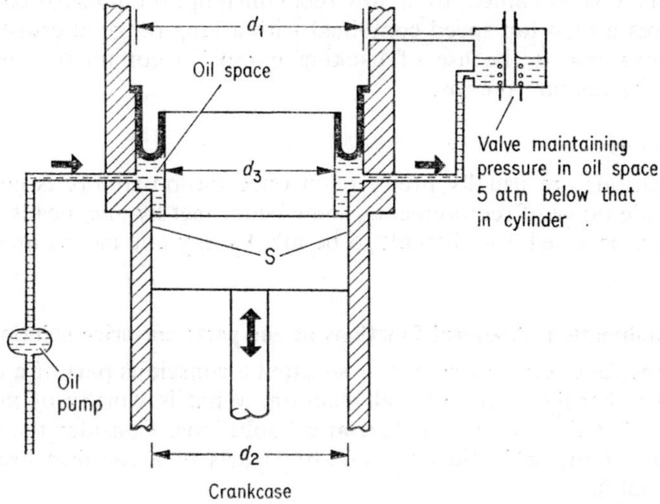

Figure 2.17 Philips rolling diaphragm hermetic seal. In this hermetic seal for pistons moving at high speeds, most of the pressure drop between the cylinder (at 100–200 atmospheres) and the crankcase occurs across the conventional seal S, only about 5 atmospheres differential pressure being carried by the rolling diaphragm. By stepping the piston diameter as shown and making the annular area between diameters d_1 and d_2 equal to that between d_2 and d_3, it is arranged that the volume of the oil space does not vary as the piston reciprocates (see section 2.12). (By kind permission of Philips N.V., Eindhoven)

Indications

Separation of function of this kind is most likely to be of value where one or both functions are very exacting or at the limits of available means, or where the parts are very large or costly.

(2) Distribution of one function between several similar parts

Obvious reasons for adopting this technique are greater reliability, greater flexibility, e.g. in coping with part load etc., using available space, and so on. A less obvious one is to save weight, as in the case of lift engines for v.t.o.l. (vertical take-off and landing) aircraft; because of the '$1\frac{1}{2}$ power' law (see section 4.6) there is a finite thrust for which the weight per unit of thrust is a minimum. The same law means that a number of small Pelton wheels may be cheaper than one large one (see section 5.5).

Another important reason for using several parts is for structural purposes, where load diffusion is concerned (see section 7.5).

(3) Division of one function between several dissimilar parts

Very often this is the key principle in a hybrid scheme, as in the hybrid power systems used in naval vessels and military vehicles, where a low-specific-weight gas turbine allied to a low-fuel-consumption internal-combustion engine gives a high top speed combined with a long range at cruising speed. Another example is the use of backing pumps in conjunction with high-performance vacuum pumps.

Indications

The indications are usually pretty plain once the possibility is considered: either a wide range of requirements, some better met by one means and some by another, or a task too difficult to be met by any one means unassisted.

(4) Combination of several functions in one part: cut-price solutions

To combine functions in one part is so often a conscious part of a designer's aims that it hardly merits special mention. What is worthy of note is the aspect of what the writer calls 'cut-price' solutions. Consider the following description of the self-adjusting actuating parts of a car disc brake (foot-operated only).

'The hydraulic fluid acts on a piston sliding in a cylinder and provided with a circumferential seal. At the forward end of its travel, the piston is gripped by a clamping device that is spring-mounted from the cylinder wall so that when the pressure is released the clamp, and hence the piston, is withdrawn a small fixed distance.'

This might be imagined to imply as many as five components besides the piston and cylinder—the seal, the clamp, means to operate the clamp, means to move the clamp forward when pressure is applied, and the spring to return the clamp after it has gripped the piston. In fact, a simple moulded sealing ring of suitable design performs all these functions by itself. The forward motion of the piston rolls the seal ring in its groove so that it is strained and tends to spring back; it also grips the piston, which will, however, slide through the seal once the rolling action has been brought to a stop by the groove. When the pressure is released, the seal ring rolls back to the unstrained or flat position, withdrawing the piston a small fixed amount.

Here we have a device whose function is relatively complicated, while its physical realization is simple and cheap. When a description of the function of a device has been framed in abstract terms, as here for the self-adjusting actuator, we should always hope for and look for such a cut-price solution. Indeed, sometimes the abstract description of a device may be so involved compared with those of other ways of doing the same job that it is only worth consideration in case a cut-price embodiment can be devised.

Indications

Cases where this technique is particularly indicated are those where the technical tasks are fairly easy or the volume of production large, military equipment and other cases where weight or space is at an especially high premium.

Auxiliary functions

It is often worth introducing an additional (auxiliary) function in order to make an exacting function easier to fill. For example, the pressure vessel in Figure 2.16 has an insulating lining and cooling pipes in the walls in order that the members performing the very exacting structural task do not have also to contend with high temperatures or high thermal stresses.

2.13 Repêchage

It has already been said in section 2.9 that rather than eliminating cautiously and finally, it is quicker and better to eliminate in a rather cavalier fashion in the first place and later to review the rejects in the improved light of the insight developed in the interim, a process here called repêchage because that is exactly what it is.

A very good test to use in the repêchage is the sort already mentioned in regard to particular options in section 2.9, to set aside the weaknesses of the scheme in question and to decide whether there is any plea to be made on its behalf, any advantage it possesses, or any circumstances in which it would be viable.

Great importance should be given to these two ideas, of rough elimination and repêchage, and of weighing advantages in the first place regardless of disadvantages, which is why they have been repeated in such a short space. The former saves time and yet reduces the chance of overlooking a good solution: the latter inspires inventions. Simple and perfect ideas are as hard to find as metallic meteorites; it is better to take drossy ore and start smelting if you need iron.

Repêchage need not be confined to your own schemes; historical devices should also be re-examined. An example of this is the celebrated I.B.M. 'golf-ball' typewriter. This incorporates all the type in hard plating on a plastic ball that is rotated in two angular degrees of freedom so as to bring any desired character opposite the paper. Now typewriters with all the type in a solid cylindrical piece, rotated and shifted along its axis to select the character, are, relatively speaking, of the highest antiquity, but have not survived for ordinary purposes. One of the chief weaknesses of the principle is the large inertia of the type, which has to be shifted bodily each time a character is typed, and the ability to make it in a material of about a tenth the density has contributed largely to making this form of typewriter viable today. The I.B.M. typewriter is full of excellent features, however, and the whole design is very refined. For example, of the two degrees of freedom the typing station must have relative to the paper, along the lines and from line to line, one is given to the paper and one to the type; this is fundamentally a better way, but not practical with the conventional typewriter, where both are given to the paper with rather clumsy results (check for yourself, which has the fundamental advantage from the inertia point of view, the sphere or the cylinder).

It is not worth attempting to list exhaustively the changed aspects or circumstances which may result in reinstating a scheme which has been provisionally eliminated. Often the designer's view about which are the greatest difficulties of the problem will change, so that slight advantages become large ones, and large disadvantages become slight. Cheaper embodiments may be devised for what at first appeared too expensive a design, perhaps involving different materials or methods of manufacture. Most often, however, it is a new idea about how the scheme in question should be worked, sometimes a new idea which has arisen in connection with some other scheme and is then seen to have even more promise for one of the arrangements eliminated in the early rounds.

2.14 Combinative ideas: general remarks

In this chapter a sketch has been given of the scope and method of working of combinative approaches. Their virtues are that they provide a way of getting to grips with difficult problems, they offer some slight guarantee against overlooking very good solutions and a near certainty of finding good ones if they are there to find, and they evolve by stages more powerful special-purpose design tools suited to the problem in hand.

They proliferate embryo schemes, often in intractable numbers, which is a very good reason for keeping tables of options and so forth as small and as simple as possible without destroying their usefulness. Reduction of the number of 'schemes' to a manageable few can usually be partly done, and occasionally completely done, by systematic elimination, and parameters descriptive of the relative difficulty of functions can sometimes be of great help. Much of the time recourse must be had to *ad hoc* methods or engineering judgement, and at this stage synthetic approaches begin to be more profitable, though the combinative table can still be used with advantage as a source of material.

Together with the insight-developing kinds of study suggested in Chapter 4, combinative approaches provide something to work on, and with, and at. This work, combined with evolutionary devices such as hybridization and redistribution of functions, can serve to build up synthesizing principles and specific design philosophies, which are the last and most powerful aids to conceptual design.

Questions

Q.2.1. It is required to design a range of high-tensile steel bolts for use in bolting together steel structures such as frames of buildings, bridges, etc. These are to be friction-grip bolts, relying on clamping the surfaces of the members joined sufficiently hard together for the friction between them to carry the applied shear loads. Typically, the shank diameter is about 30 mm, and to guarantee sufficient residual load in the bolts it is necessary to take them into yield. Clearly, it is important to be able to ensure that the bolts are correctly tightened. Make a combinative study of this problem.

Q.2.2. Study the problem of powering deep-sea exploratory submarines.
Q.2.3. Consider the type of garden motor-mower which has a bladed cylinder rotating against a stationary blade and a driven roller at the rear—a typical small mower of the cylinder type, in fact. What arrangements of clutches are possible, and what are their merits?
Q.2.4. Make a classification of valve-operating gear for four-stroke petrol engines for private cars, from crankshaft to valve stem.
Q.2.5. The strain in the bore of aircraft gas turbine discs due to centrifugal loads is of the order of 0.003, so that a disc on a shaft of 100 mm diameter will be loose by 0.3 mm when running if nothing is done about it. List as many practicable methods of centring such discs as you can think of, and try to find a system of classification. Make your approach as abstract and exhaustive as you can.
Q.2.6. Analyse the problem of fore-and-aft stability of prams (baby carriages), and list the means of solution in as general and abstract a form as seems useful. Note any practical examples.
Q.2.7. Make a functional analysis and a table of options for kitchen tin-openers for opening cylindrical tins with the usual rolled rim joint.

Notes

For this sort of problem, complete answers are not possible. The following notes are offered.
N.2.1. The dominant functions are 'enabling tightening to the correct load' and 'ensuring tightening'. A torque spanner enables tightening to the correct load but does not ensure it; there is a premium on a bolt design which can be quickly inspected for correct tightening afterwards. It is valuable to consider as separate means of tightening imposing a given load (as with a torque spanner) or imposing a given distortion (as by turning $\frac{3}{4}$ turn after the nut tightens). Means of ensuring tightening may be by observing a given distortion. Then the distortion may be of the bolt shank itself, of the bolt-head, of the nut, of the washer or of some component used in tightening. (In the Torshear bolt this is an extension of the bolt shank which is used to react the nut-tightening torque; tightening continues until this extension is twisted off, leaving clear evidence of correct tightening.) In the GKN load-indicating bolt, the square head is undercut on its bearing face except for four triangular areas at the corners; under the correct tightening load the head partially collapses, reducing the gap between the undercut part of the underside of the bolt head and the parts joined. Gauging of this gap ensures the correct loading.
N.2.2. The parametric plot of Figure 2.18, which shows the viable regions for types of power supply for spacecraft, has some bearing on this problem (but see section 9.6).

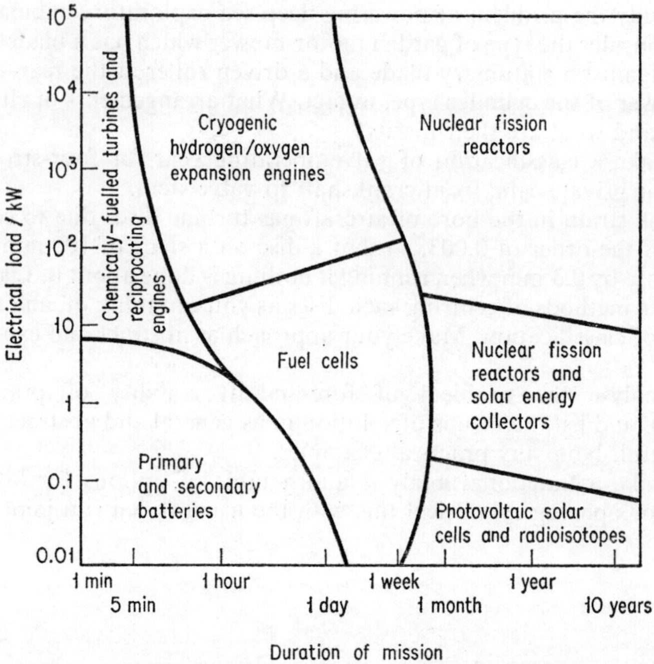

Figure 2.18 Plot of viable power sources for use in space. (By kind permission of Dr. B. C. Lindley and the *Journal of the British Interplanetary Society*)

N.2.3. Basically, there are three shafts—engine (E), roller (R) and cutter (C) —which can be connected with one or two clutches in the following ways, where X = clutch and ( ) denotes a solid shaft

One clutch		*Two clutches*
1 E X (CR)	both have merits, but the	4 E X C X R
2 (EC) X R	second has severe snags also	5 C X E X R
3 (ER) X C	this seems useless	6 E X R X C

Note that while 5 is more versatile than 4, it has the disadvantage that to declutch the engine entirely means operating two controls instead of merely one. The battle is between 1 and 4.

N.2.4. Some points of classification are:

camshaft drive—bevels, helicals, spurs, chain, toothed belt, multiple eccentrics,
rocker arms or direct operation,
valve closure—by springs or 'positive' (desmodromic).

N.2.5. See section 7.4 and Figure 7.14.

N.2.6. High stability is safe, but makes it difficult to negotiate kerbs, because the pram is more difficult to tilt up to get the front wheels on the pavement. Radical solutions may be classified as:

(1) Low initial stability with high terminal stability, i.e. just prior to oversetting,
(2) Variable-geometry prams (the variation probably being achieved by the same operation as setting the brake, since brake-off/high-stability and brake-on/low-stability conditions are never needed).

Both solutions have been or are available.

N.2.7. For example,

Function	*Options*
Cutting	Wheel, knife
Driving cutter	Rollers, cutting wheel, toothed wheels, hand
Source of power	Motor, hand
Guiding cutter	Rim of tin, preset radius arm, hand
Retaining lid	None, magnet, sucker

(the first line contains an arbitrary decision)

3 Optimization

3.1 Introduction

In the last chapter we considered the problem of how to select the best arrangement or combination of means for performing a given function. The variations under consideration were of kind, not of degree. But even when all decisions of kind have been made, so that all the qualitative aspects are settled, there remains the easier problem of fixing all the quantities involved. The solution of this problem we shall call 'optimization'.

Before considering optimization generally four examples will be examined to fix ideas and illustrate important points.

3.2 Making capital and running costs commensurate

In all our optimizations the ultimate object will be minimum cost, though sometimes a more immediate object like minimum weight may be substituted temporarily. In dealing with the structural aspects of aircraft engines, for example, it is almost possible to equate minimum cost with minimum weight, and the latter criterion is easier to work with.

Costs are of many kinds and given in different terms, and before their total can be minimized they must be put on a common footing. Associated with a prime mover, say, there may be a first cost, in pounds, a fuel cost, in pounds per tonne, a maintenance cost in pounds per 10 000 hours and so on, and the total cost may be expressed in pounds per kilowatt hour, or pounds per year. To combine the separate costs is generally simple enough but it depends upon other data, such as operating hours per year. In particular, capital cost depends upon many items, notably, interest and taxation rules, and is treated in practice in various ways, some of which are clearly inappropriate. For the present purpose, where it is necessary to make capital and running costs commensurate, it will be assumed that this is done by reducing capital cost to a fixed annual charge sufficient to pay back the capital over the life contemplated while yielding a stipulated percentage on the capital still outstanding. This is exactly what happens under the usual form of mortgage, where the householder pays an annual sum consisting of interest on the borrowed money outstanding and an element of repayment; the sum is the same every year, but is fixed at such a level that the money is all repaid by the end of a stipulated period. If p is the interest, P the amount borrowed, and n

the number of years for which the mortgage is to run, the fixed annual payment is kP, where

$$k = \frac{p}{1-(1+p)^{-n}}$$

(check that this expression tends to the correct limit as $p \to 0$).

For example, if $n = 18$, $p = 0.10$ (i.e. 10 per cent), $k = 0.122$, so the householder pays £122 per annum for £1000 borrowed. The case of an industrial investment is similar, except that p is the return expected and n the expected life.

In many projects, where the return and costs are not fixed over the life, a more complicated treatment is needed, but it can be based on exactly the same idea (Discounted Cash Flow: see [42]). Such complications and those introduced by taxation and capital gearing are beyond the present scope.

3.3 Optimum speed of a tanker

We shall study in a very simplified form the question of the optimum speed for an oil tanker of given size, forming part of a fleet sufficiently large for the repercussions of this decision on it to be negligible.

We shall minimize the costs per tonne-kilometre of cargo carried. The *annual* costs can be divided into two categories, those which are independent of speed (total a) and those which are dependent on speed (total b). The most important items are:

a: cost of hull, fittings, and some machinery, crew costs,
b: some of main engine costs, some machinery, fuel.

The drag of the ship is roughly proportional to v^2, where v is the speed, and so the power required is proportional to v^3. Thus the total annual costs can be written in the form

$$a + b\frac{v^3}{v_0^3} \tag{3.1}$$

where v_0 is some reference value of v for which the costs b are calculated. But the tonne-kilometres per year are proportional to v, so that the costs per tonne-kilometre are proportional to

$$\frac{\textit{annual costs}}{v} = \frac{a}{v} + \frac{bv^2}{v_0^3} \tag{3.2}$$

The problem, then, is to choose v so as to make expression 3.2 a minimum, which is done by putting

$$\frac{\mathrm{d}}{\mathrm{d}v}\left(\frac{a}{v} + \frac{bv^2}{v_0^3}\right) = 0$$

or

$$-\frac{a}{v^2} + \frac{2bv}{v_0^3} = 0$$

so that
$$v = \left(\frac{a}{2b}\right)^{\frac{1}{3}} v_0 \tag{3.3}$$

the required optimum or *cheapest* speed. Substituting this value of v in expression 3.2, it takes the form

$$\frac{a}{v_0}\left(\frac{2b}{a}\right)^{\frac{1}{3}} + \frac{b}{v_0}\left(\frac{a}{2b}\right)^{\frac{2}{3}} = (2+1)\frac{a^{\frac{2}{3}}b^{\frac{1}{3}}}{2^{\frac{2}{3}}v_0} \tag{3.4}$$

i.e. the speed-independent costs are twice as great as the speed-dependent ones at the optimum speed.

Suppose now a special tanker is built which is necessarily more expensive to build than the ordinary sort, but has ordinary propulsive machinery. If we held the speed at the customary value for the size of ship in question, the *a*-type, or speed-independent costs, would be greatly increased, leaving the *b*-type, or speed-dependent costs, unaltered, and the 2:1 ratio of the two types would be upset. From equation 3.3, we see that the speed should be increased as the third root of the *a*-type costs.

The liquid natural gas (L.N.G.) tankers mentioned in the last chapter exemplify this. Because of their expensive hull constructions and the high cost of the tanks, insulation and cargo-handling machinery, the *a*-type costs are increased by a factor of about 2 for a given hull size. If the speed were only the same as that of an ordinary tanker, costs per kilometre would go up from

2 units (*a*-type) + 1 unit (*b*-type)
to 4 units (*a*-type) + 1 unit (*b*-type)

i.e. from a total of 3 units to a total of 5 units.

However, from equation 3.3, we should put the speed up by a factor of $2^{\frac{1}{3}}$, and from expression 3.4 the costs become

$2 \times 2^{\frac{2}{3}}$ units (*a*-type) + $2^{\frac{2}{3}}$ units (*b*-type)

a total of $3 \times 2^{\frac{2}{3}} = 4.76$ units.

This illustrates a general principle of great importance, that any change in circumstances of a design ought to be followed through in all its consequences. The speed of an L.N.G. tanker is about 17.2 knots, as against the 13.5 knots or so of a tanker of the same size, an increase of about 27 per cent ($2^{\frac{1}{3}} = 1.26$). However, there are other factors acting in favour of a higher speed for L.N.G. tankers, besides high capital cost.

Figure 3.1 shows the variation of L.N.G. transport costs with speed. Notice that in the above simple calculation, an increase in speed of 26 per cent reduced costs by less than 5 per cent (5 to 4.76 units) and the curves of Figure 3.1 show the same flatness.

It may be objected that the foregoing analysis is quite inadequate, that the representation of drag by a v^2 term is inaccurate, and so forth. This is so, but as a simple illustration it is valid enough. The two to one ratio at the optimum speed of the two large categories of cost may not be very precise, but it does show the kind of logic characteristic of situations of this sort.

Three points should be noted:

(1) Curves such as those in Figure 3.1 tell much more than just where the optimum is; they show how rapidly the cost rises as the optimum is departed from (see section 3.10).

(2) The conditions determine the optimum; if the tanker problem is worked out with the tonne-kilometres per annum fixed instead of hull size, a slightly different balance is found, with a and b type costs in the ratio 7:3 instead of

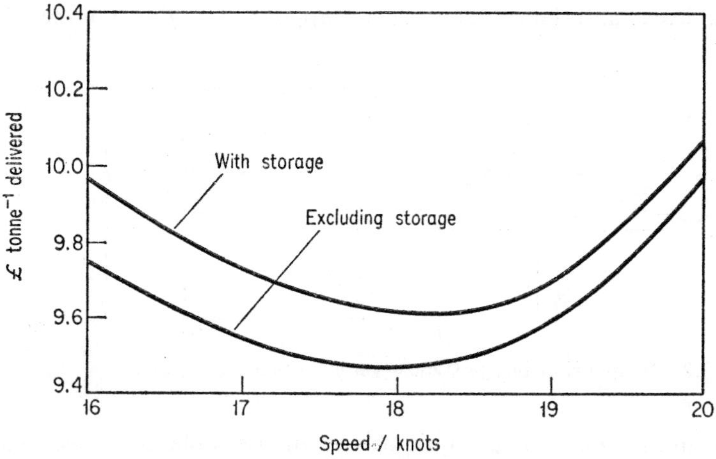

Figure 3.1 Speed of L.N.G. tanker

2:1. Note that in this case M, the tonnage, is variable as well as v, the speed, but that Mv must be kept constant. The drag is proportional to area times v^2 or $M^{\frac{2}{3}}v^2$.

(3) Above all, the consequences of changes must be followed through in all their implications, if disadvantageous ones are to be mitigated and advantageous ones exploited to the full.

3.4 The optimization of the sag/span ratio of a suspension bridge

Let us look very briefly at an example rather similar to the last before proceeding to a more difficult one. Figure 3.2 is a diagram of a suspension bridge of span s and central sag h. Suppose the deck to be of constant weight per unit length w, and to have side spans of length $s/2$, so that each tower carries a downward load ws (neglecting the weight of the cables and hangers). If $\frac{1}{2}T$ is the tension in each of the two cables, the vertical component of T must be $\frac{1}{2}ws$. Since the cable hangs in a parabola, the tangents PR, PS to it at the tower tops intersect at point P, h below the centre of the cable.

The vertical component of T is $T \sin \theta = \frac{1}{2} ws$. In practice, the sag is small, so

$$\sin \theta \simeq \frac{2h}{\frac{1}{2}s} = \frac{4h}{s}$$

and
$$T \simeq \frac{ws^2}{8h}$$

The tension in the cables, and so their cross sectional area, is thus inversely proportional to h. If we make the towers taller and h greater, the towers become dearer and the cables cheaper. Suppose f_c, c_c, f_t, c_t are the allowable

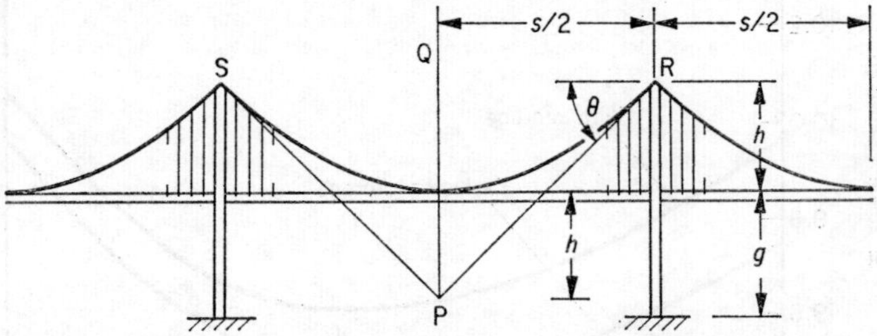

Figure 3.2 Suspension-bridge statics (vertical scale exaggerated)

stresses and *in situ* costs per unit volume of the cable and tower materials, respectively, we can write down a rough total cost of cables and towers like this. The combined cross-sectional area of the cables is

$$\frac{T}{f_c} \simeq \frac{ws^2}{8hf_c}$$

their length is roughly $2s$, and their cost is approximately

$$\frac{ws^2}{8hf_c} \times 2s \times c_c = \frac{ws^3 c_c}{4hf_c}$$

The cross-sectional area of each tower is ws/f_t, its height is $(h + g)$ (see Figure 3.2) and its cost is

$$\frac{ws}{f_t}(h + g)c_t$$

The total cost C_T of two towers and the cables is given by

$$C_T = 2ws(h + g)\frac{c_t}{f_t} + \frac{ws^3}{4h}\frac{c_c}{f_c} \tag{3.5}$$

To minimize C_T we put

$$\frac{dC_T}{dh} = 0 = 2ws\frac{c_t}{f_t} - \frac{ws^3}{4h^2}\frac{c_c}{f_c}$$

$$\text{giving } \left(\frac{h}{s}\right) \text{ optimum} = \left(\frac{1}{8}\frac{c_c}{f_c}\frac{f_t}{c_t}\right)^{\frac{1}{2}} \tag{3.6}$$

Putting some very rough values in equation 3.6, say, $f_c = 600 \text{ N mm}^{-2}$, $f_t = 120 \text{ N mm}^{-2}$, $c_c = £1200 \text{ m}^{-3}$ and $c_t = £2400 \text{ m}^{-3}$, gives $h/s = 1/9$, which corresponds roughly with present practice. Notice that c_t will be very high here (more than four times the raw-material cost) because of the costs of prefabrication, transport to the site and erection, the last two being a function of weight. Also c_c is only the marginal cost of adding a cubic metre more cable. Although high-yield steel will clearly pay its way for the towers, f_t for this purpose is the part of the allowable stress which can be allotted to taking direct load, since bending also occurs (see section 6.10).

Substitution from equation 3.6 in equation 3.5 shows that in the optimized case, the cost of the cables is equal to the cost of the h or variable part of the towers (that above the lowest point of the cables; see Figure 3.2).

The simple treatment of equation 3.5 does not adequately represent the complexity of the problem. The weight of the cables contributes substantially to the load carried by the towers; increasing the height of the towers adds to their cost more than proportionately; the costs of the hangers by which the deck is suspended and the anchors at the ends of the cables need incorporation (the former will behave like tower cost, the latter, like cable cost); above all, the important problem of stiffness needs taking into account.

Stiffness in the vertical direction comes from two sources—the deck structure and the cables. The deck can be stiffened by attaching trusses, or it can form the top of a box girder. The cables are able to provide stiffness because of the heavy loads in them; to bend the deck it is necessary to alter the polygonal curve of the cables, and this requires forces which will be in direct proportion to the tensions in them. The lower the sag/span ratio, the higher the tensions in the cables and the less the need of a stiffening truss or stiff deck.

All this can be seen in Figure 3.3 where the problem we have treated very roughly has been done properly. It will be seen how the cable cost varies very nearly inversely as the sag, and the anchorages vary nearly in the same way but with a larger fixed term. The towers, however, increase in price somewhat more rapidly than proportionately to the sag.

3.5 Optimization with more than one degree of freedom—heat exchanger

The two examples considered so far involve only one degree of freedom of choice—in the first case we could only vary the speed of the ship v, and in the second case we could only vary the sag/span ratio, h/s. In many practical cases we have more than one degree of freedom of choice, more than one thing we can vary to improve our criterion, here taken to be low cost.

Consider the case of a heat exchanger: we would like this to have zero capital cost, to produce no pressure drops, and to have a zero minimum temperature difference between the two sides. Compared with this imaginary device, a real exchanger involves three items of cost:

(1) capital cost,
(2) costs associated with the pressure drops,
(3) costs associated with the temperature difference between the two sides.

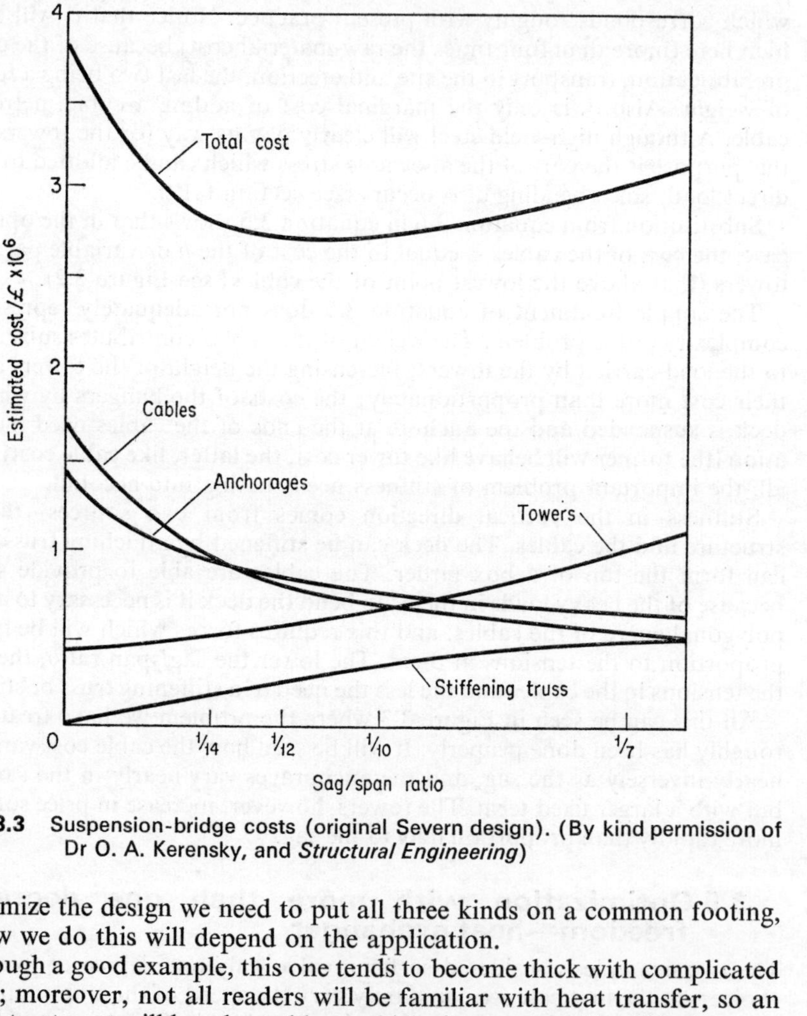

Figure 3.3 Suspension-bridge costs (original Severn design). (By kind permission of Dr O. A. Kerensky, and *Structural Engineering*)

To optimize the design we need to put all three kinds on a common footing, and how we do this will depend on the application.

Although a good example, this one tends to become thick with complicated algebra; moreover, not all readers will be familiar with heat transfer, so an unusual treatment will be adopted here which circumvents this difficulty.

To simplify the problem as far as possible, let us study only the tube side of a counter-flow heat exchanger in which a fluid of substantially constant properties is heated by a tube wall everywhere δT hotter than itself. The pressure drop in the fluid is Δp, and occurs entirely in the tubes. The total area of tube wall is A. The total cost C_T is put in the form

$$C_T = k_1\alpha + k_2\pi + k_3\tau \qquad (3.7)$$

where k_1, k_2 and k_3 are constants,

$$\alpha = \frac{A}{A_0}, \pi = \frac{\Delta p}{p} \text{ and } \tau = \frac{\delta T}{T}$$

and A_0, p and T are a suitable area and the pressure and temperature at entry to the heat exchanger respectively. The three terms on the right-hand side of equation 3.7 are the capital cost, the cost of the pressure drop, and the cost of the temperature differential respectively, all reduced to some common basis such as annual cost, total cost over a given life, cost per kilogramme of fluid passing etc. (see section 3.2).

In general we can reduce any of these terms at the expense of either of the others. By keeping A constant but making the tubes fewer in number and longer, we can increase the speed of flow and so decrease τ, but at the same time π will be increased. We can reduce both π and τ by increasing A in a suitable fashion, and so on.

Let α_0, π_0, τ_0 be the values of α, π, τ for some trial design. Consider the effect of increasing the number of tubes by a factor x and their length by a factor y. Then the area of tube-wall will be increased in the ratio xy, and α becomes $\alpha_0 xy$. The velocity of the fluid is reduced in the ratio $1/x$ since it is divided between x times as many tubes and the pressure drop per unit length of tube is proportional to the velocity to the power of 1.8 roughly, so that π becomes

$$\pi_0 x^{-1.8} y$$

Finally, the rate of heat transfer per unit area of tube wall is roughly proportional to the velocity to the power of 0.8, the area is up in the ratio xy and τ becomes

$$\frac{\tau_0 x^{0.8}}{xy} = \tau_0 x^{-0.2} y^{-1}$$

For a general design scaled from the trial one by the factors x, y,

$$C_T = k_1\alpha_0 xy + k_2\pi_0 x^{-1.8} y + k_3\tau_0 x^{-0.2} y^{-1} \qquad (3.8)$$

Now C_T is minimized with respect to x, y by putting

$$\frac{\partial C_T}{\partial x} = \frac{\partial C_T}{\partial y} = 0 \text{ (see section 3.10) giving}$$

$$k_1\alpha_0 y - 1.8k_2\pi_0 x^{-2.8} y - 0.2k_3\tau_0 x^{-1.2} y^{-1} = 0 \qquad (3.9)$$

$$k_1\alpha_0 x + k_2\pi_0 x^{-1.8} - k_3\tau_0 x^{-0.2} y^{-2} = 0 \qquad (3.10)$$

Multiplying equation 3.9 by x and equation 3.10 by y and substracting, an expression for y in terms of x is found:

$$y = \left(\frac{2}{7}\right)^{\frac{1}{2}} \left(\frac{k_3\tau_0}{k_2\pi_0}\right)^{\frac{1}{2}} x^{0.8} \tag{3.11}$$

Substituting for y in equation 3.9, we find

$$x = 2.5^{\frac{5}{14}} \left(\frac{k_2\pi_0}{k_1\alpha_0}\right)^{\frac{5}{14}} \tag{3.12}$$

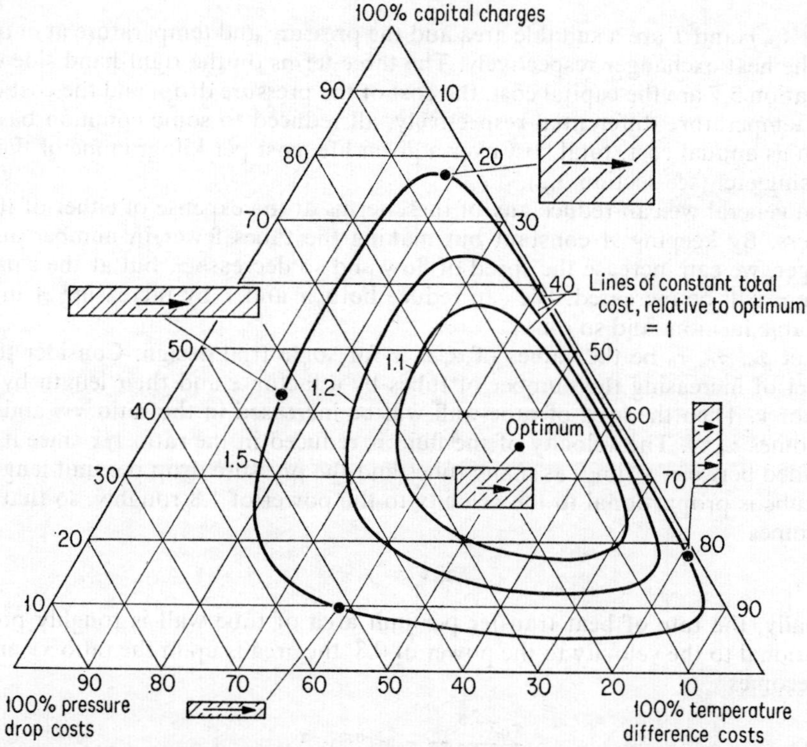

Figure 3.4 Total costs of heat exchanger. The cross hatched rectangles show the relative sizes of different designs

Substituting for y and x in equation 3.8 we find for the minimum value of C_T

$$C_T\,(\text{min}) = 2.7 \left(\frac{5}{14} + \frac{2}{14} + \frac{7}{14}\right) (k_1\alpha_0)^{\frac{5}{14}} (k_2\pi_0)^{\frac{2}{14}} (k_3T_0)^{\frac{7}{14}} \tag{3.13}$$

where the fractions have been kept to show the division of C_T (min) into 5/14 attributable to capital cost, 1/7 attributable to pressure losses, and 1/2 due to temperature difference.

Figure 3.4 shows that large departures from the ideal distribution between capital, pressure drop and temperature difference costs are possible without much loss—for example, when all three are equal the total cost is about 10 per cent above the minimum.

There are many ways in which this analysis could be invalid in a given case: to mention a few, there may be a fundamental temperature mismatch (section 5.7) the cost of end fittings has been ignored and the flow may tend to move out of the turbulent regime. The first two complicate the problem; the last can change its character, with the optimum at a discontinuity and not a turning value (section 3.10).

The treatment of heat exchangers of all kinds, with changes of phase, variation of fluid properties, different configurations and so forth, will not be attempted here, but a few simple general observations are made in the next two sections.

It should be noted that the trick of using scale factors x and y separates the difficulties of the heat transfer aspect from those of the optimization; the full value of this device is best appreciated by attempting the same problem without its aid.

3.6 Putting a price on the heat-exchanger performance

We have already noted the need to make capital and running costs commensurate. We now study the problem of putting a cost on performance figures

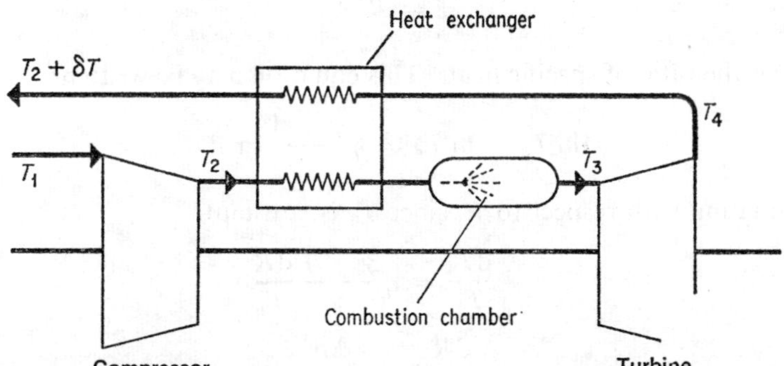

Figure 3.5 Gas-turbine flow diagram

such as π and τ, i.e. of giving values to k_1 and k_2 in equation 3.7. The procedure will be illustrated by reference to a simple open-cycle gas turbine, an example that brings out several important points.

Figure 3.5 is a flow diagram for a gas turbine in which the air leaving the compressor at temperature T_2 is heated in a counterflow heat exchanger by the exhaust gases. As the products of mass flow and specific heat for the air and the exhaust gases are nearly equal, an ideal heat exchanger would raise the air temperature to T_4, the temperature of the gases leaving the turbine. If T_3 is the turbine inlet temperature, the temperature rise through the combustion chamber would be $T_3 - T_4$. From all points of view, it is desirable to keep T_3

fixed at the highest value the blade material will permit. With a real heat exchanger the air will only be raised to $T_4 - \delta T$, where δT is the temperature difference in the heat exchanger. Since T_3 is fixed, the temperature rise through the combustion chamber will be $T_3 - T_4 + \delta T$. To achieve this, the fuel flow must be increased above that which would be necessary with an ideal exchanger, roughly in the ratio

$$\frac{T_3 - T_4 + \delta T}{T_3 - T_4},$$

and it is a straightforward matter to put a cost on this additional fuel consumption.

It is not so simple to put a cost on Δp, the pressure drop through the heat-exchanger side. The effect of Δp is to reduce the pressure ratio R over the turbine, reducing the temperature drop $T_3 - T_4$ (since T_3 is held constant). The work done in the turbine per unit mass is reduced, and the output power is reduced by the same amount since the power required to drive the compressor is unaltered. Presuming the output power to be specified, we must increase the mass flow.

To estimate the decrease in $T_3 - T_4$, it is reasonable to assume that the polytropic efficiency of the turbine remains unchanged at η, say. Then

$$\frac{T_3}{T_4} = R^{\eta\frac{\gamma-1}{\gamma}}$$

where γ is the ratio of specific heats. This equation may be written

$$\ln T_3 - \ln T_4 = \eta \frac{\gamma - 1}{\gamma} \ln R$$

Differentiating with respect to R, since T_3 is constant,

$$-\frac{\mathrm{d}T_4}{T_4} = \eta \frac{\gamma - 1}{\gamma} \frac{\mathrm{d}R}{R}$$

and
$$\frac{\mathrm{d}R}{R} \simeq -\frac{\Delta p}{p}$$

where p is the mean pressure in the side of the heat exchanger being considered.

If $\Delta p/p$ increases by 1 per cent in a case in which $T_4 = 0.7\,T_3$, $\eta = 0.84$ and $\gamma = 1.4$,

$$\frac{\mathrm{d}T_4}{T_4} = \frac{0.84(1.4 - 1)}{1.4} \times 1\% = 0.24\%$$

and the temperature drop through the turbine decreases by a fraction

$$\frac{T_4}{T_3 - T_4} \times \frac{\mathrm{d}T_4}{T_4} = \frac{0.7}{0.3} \times 0.24\% = 0.56\%$$

This is the drop in turbine shaft power. The output power may be 45 per cent of the turbine power, perhaps, the remainder driving the compressor. In such a case the drop in output power is

$$\frac{100}{45} \times 0.56\% = 1.24\%$$

To restore this output power drop requires the same increase of mass flow.

The rise in fuel consumption will not be quite as much as 1.24 per cent because the rise in T_4 produces an increase in the temperature of the air leaving the high-pressure side of the heat exchanger; as a result the temperature rise in the combustion chamber is less. In a typical case, the increase in fuel consumption might be about 0.9 per cent.

Having found that, if the power output is to be maintained, an additional pressure drop of 1 per cent through the high-pressure side of the heat exchanger means an increase of about 0.9 per cent in fuel consumption and about 1.2 per cent in mass flow (and hence of machine size generally), we can put a cost on these penalties, say, in new pence per kilowatt hour. Naturally, the pressure drop and temperature difference in the other side of the heat exchanger must be included in a full treatment.

The approach we have used to find the effects of a small increase of pressure drop is of a kind called here a *differential study*. It is quick, accurate and cheap, and excellent for developing insight (section 4.3).

3.7 Variation of costs with application

As with the tanker speed problem, it is important to recognize that a change in any k in equation 3.7 will not affect the relative sizes of the terms $k_1\alpha$, $k_2\pi$ and $k_3\tau$ in the optimized case, but will affect the relative sizes of α, π and τ. For example, in a prime mover, τ often involves only higher fuel consumption, whereas π involves both higher fuel consumption and higher first cost. In a mechanical refrigerator, where both π and τ demand increased inputs of mechanical work, they are on a more equal footing and will tend to appear in the proportions that fundamental thermodynamics suggests. For a refrigerator, we can almost write, instead of equation 3.7,

$$C_T = k_1\alpha + k_4\Delta s$$

where Δs is the increase in specific entropy through the exchanger (since there is no net change of enthalpy in the exchanger, the increase in entropy in it multiplied by the ambient temperature is the lost work). We must expect the ratio of π to τ to be higher in the exchangers of an air separation plant, say, than in those of a gas turbine.

3.8 Further aspects of heat-exchanger optimization

Before leaving the heat-exchanger problem, there are two important points to be mentioned briefly.

The possibility of varying element size has been considered, but usually this aspect can be settled separately. Sometimes the answer is, the smaller the diameter of the tubes the better, within the limits set by fouling. Sometimes the cost of end joints is dominant, or perhaps limits are imposed on the fluid velocity.

The question of element type is also important. In some cases plain tubes, or perhaps tubes with integral fins, are the only acceptable solution, limitations being imposed by pressure, thermal shock etc. In other cases a wide choice of elements are available. On the whole, the more elaborate forms of element increase both the local heat-transfer rate and the resistance to flow, and the balance of these conflicting effects may swing either way according to the relative sizes of the *k*'s and the properties of the fluids and the solid involved [11].

3.9 An elementary programming problem

Consider four parts W, X, Y, Z, which are used together in two alternative combinations,

one X, two Y's and one Z: arrangement A
or one X, one Y and one W: arrangement B

In each combination there is one critical fit involving several toleranced dimensions, one from each part; we shall use the symbols W, X, Y, Z to denote the tolerances on these dimensions, and W_0, X_0, Y_0, Z_0 to denote large reference values of the tolerances. We shall also write

$$W = W_0 - w, X = X_0 - x, \text{ etc.}$$

so that w, for example, represents the amount we have tightened the tolerance from the rather open reference value W_0. For each complete set of parts to be produced, suppose the cost of this reduction w in tolerance is $2w$ pence, i.e. w costs 2p. a unit. Suppose also that x costs 6p. a unit, y costs 9p. a unit and z costs 2p. a unit. Then the total cost of tightening all the tolerances by w, x, y, z, respectively, is

$$C_T = 2w + 6x + 9y + 2z \text{ p.}$$

To fix ideas, the units might be thousandths of inches. If the tolerances W_0, X_0, Y_0, Z_0 are 1.5, 1.0, 1.0 and 1.5 units, respectively, then the range of fit that they would produce with arrangement A as between maximum and minimum metal conditions would be

$$X_0 + 2Y_0 + Z_0 = 4.5$$

and with arrangement B

$$X_0 + Y_0 + W_0 = 3.5$$

Suppose now that for functional reasons these ranges of fit are too great, and we need to reduce them to 2.5 and 2.0, respectively, so that

$$X + 2Y + Z = X_0 - x + 2(Y_0 - y) + Z_0 - z$$
$$= (4.5 - x - 2y - z) = 2.5$$

or $x + 2y + z = 2.0$

and similarly, $X + Y + W = 2.0$, reducing to $x + y + w = 1.5$

We require to satisfy these two equations with values of w, x, y, z that make C_T as small as possible; to make a reasonable problem, however, we must limit the values of w, x, y, z. Clearly, $W = W_0 - w$ cannot be indefinitely small; let us limit it to 0.8, so that $w \leqslant 0.7$. Let us impose also the limits $X \geqslant 0.3$, $Y \geqslant 0.5$, $Z \geqslant 0.5$, and write down the entire problem in formal mathematics:

Minimize

$$C_T = 2w + 6x + 9y + 2z$$

subject to the equations

$$\left.\begin{aligned} x + 2y + z = 2.0 \\ x + y + w = 1.5 \end{aligned}\right\} \quad (3.14)$$

and the inequalities

$$w \leqslant 0.7, x \leqslant 0.7, y \leqslant 0.5, z \leqslant 1.0$$

This is a very simple problem in *linear programming* (see next section).

By means of the equations, we eliminate w and z from the expression for C_T and the inequalities, giving

$$C_T = 7.0 + 2x + 3y \quad (3.15)$$

$$\left.\begin{aligned} w = 1.5 - x - y \leqslant 0.7, x + y \geqslant 0.8 \\ z = 2.0 - x - 2y \leqslant 1.0, x + 2y \geqslant 1.0 \end{aligned}\right\} \quad (3.16)$$

There are now four inequalities governing x and y, two in 3.14 and two in 3.16.

On an x, y plot each of these inequalities can be represented by a straight line (see Figure 3.6) and all four are satisfied by any point in the interior of the quadrilateral so formed, and only by such points. The region in which all the imposed conditions or 'constraints' are satisfied is called the feasible area, and in this case it is a quadrilateral (ABCD in Figure 3.6). (Notice that if the straight line AD were much lower in Figure 3.6, it would not be an effective constraint and the feasible area would be a triangle.) The problem reduces to finding the point in the feasible area for which C_T is least.

A constant value of C_T corresponds to a line of slope $- 2/3$ (see equation 3.15). Thus we have only to draw the line of this slope and nearest to the

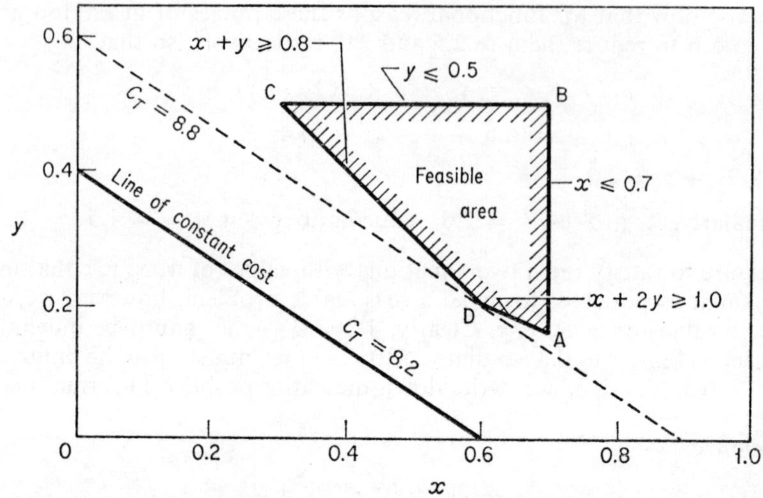

Figure 3.6 Tolerancing problem

origin which cuts the feasible area. Thus D ($x = 0.6$, $y = 0.2$) is the required solution, giving $w = 0.7$, $z = 1.0$ and $C_T = 8.8$.

Serious linear programming problems are much more elaborate, and require special methods for their solution.

There is a useful approach to a very easy problem like this one, the task/cost viewpoint. The task is to effect a total reduction of range of fit of

$$(4.5 + 3.5) - (2.5 + 2.0) = 3.5$$

Now a unit of x costs 6p., and reduces both ranges of fit (i.e. in arrangements A and B), a task/cost ratio of $\frac{2}{6} = \frac{1}{3}\ 10^{-3}$ in/p. A unit of y costs 9p. and reduces one range of fit by 2 and the other by 1, giving the same task/cost ratio, and so on. We have

	task/cost ratio, 10^{-3} in/p.
w	$\frac{1}{2}$
x	$\frac{1}{3}$
y	$\frac{1}{3}$
z	$\frac{1}{2}$

Clearly, w and z are the best value for money and we 'buy' all we can of these, 0.7 of w and 1.0 of z for 3.4p. The remaining $(3.5 - 1.7)$ or 1.8 units we have to take in x and y, at a cost of $1.8/\frac{1}{3} = 5.4$p., giving a minimum C_T of 8.8p.

3.10 Classification of optimization problems and methods of solution

Having studied four cases of optimization, the simple ones of the speed of the tanker and the sag/span ratio of the suspension bridge, the two degree of freedom example of the heat exchanger and a simple programming problem, let us glance briefly at the whole field.

Problems can be classified in two ways:

(a) by the number of degrees of freedom, and
(b) by the kind of maximum.

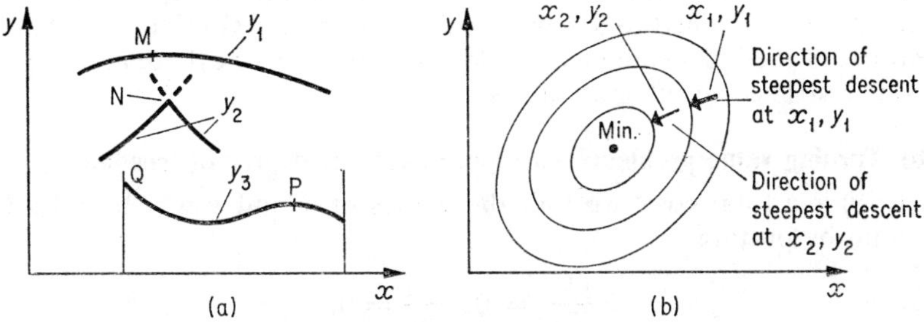

Figure 3.7 (a) Kinds of maximum (b) Method of steepest descents

Figure 3.7 shows three functions of x, y_1, y_2 and y_3: y_1 has a maximum of the classical sort at M, found by putting $\mathrm{d}y_1/\mathrm{d}x = 0$; y_2 has a maximum at N, but $\mathrm{d}y_2/\mathrm{d}x$ is not zero there; finally, y_3 has a maximum of the classical kind at P, but the maximum value in the range shown occurs at Q.

Problems where a maximum such as M or P is sought are called turning value problems; problems where a maximum such as N or Q is sought are called programming problems. In practical situations it is not always possible to tell beforehand which kind of maximum will prove to be concerned (see Question 3.5).

In principle, programming problems are much the easier kind to solve, but in practice they generally have such large numbers of degrees of freedom as to make them very difficult. Their solution is beyond our scope here, but there is an abundance of literature [12, 13]. A few remarks on the solution of turning value problems will be made.

(a) Turning value problems with one degree of freedom

If $C_T(x)$ is a total cost, the minimum is found by putting $\mathrm{d}C_T/\mathrm{d}x = 0$.

It is seldom if ever necessary to use the analytical tests, $\mathrm{d}^2C_T/\mathrm{d}x^2 > 0$ for a minimum, $\mathrm{d}^2C_T/\mathrm{d}x^2 < 0$ for a maximum because the sense of the problem will serve instead. In the case of the tanker,

$$C_T = \frac{a}{v} + bv^2$$

tends to infinity both as v approaches 0 and as v approaches infinity, for very real reasons, so that the intervening turning value must be a minimum.

If C_T cannot usefully be differentiated or the resulting equation readily solved, it may be best to evaluate C_T at a number of points. Special rules have been found for doing this economically (e.g. Fibonacci search), but these are rarely of use. In the problem of section 6.4, the disc-brake caliper, it is more useful to plot the function concerned than just to find its minimum, and this is often the case.

When the value of the independent variable is restricted to certain fixed values, integers perhaps, the turning value is found as though this restriction did not exist and then the nearest permitted values of the independent variable on either side examined to see which gives the practical minimum. Thus, in Question 3.8 the independent variable is the number of cylinders in a pump, which is restricted to integral values.

(b) Turning value problems with more than one degree of freedom

If $C_T(x,y)$ is a total cost, we find the values of x and y which make C_T a minimum by putting

$$\frac{\partial C_T}{\partial x} = 0, \frac{\partial C_T}{\partial y} = 0,$$

and testing the turning values so found, if necessary, by the tests

(1) $$\left(\frac{\partial^2 C_T}{\partial x \partial y}\right)^2 < \frac{\partial^2 C_T}{\partial x^2} \cdot \frac{\partial^2 C_T}{\partial y^2}$$

for a minimum or maximum, as against a saddle

(2) $$\frac{\partial^2 C_T}{\partial x^2} > 0, \frac{\partial^2 C_T}{\partial y^2} > 0 \text{ for a minimum}$$

(3) $$\frac{\partial^2 C_T}{\partial x^2} < 0, \frac{\partial^2 C_T}{\partial y^2} < 0 \text{ for a maximum}$$

(either half of conditions (2) and (3) implies the other half if (1) is established).

Only the simplest problems can be treated in this way, and more generally useful techniques are a specialized topic which cannot be dealt with here. The most celebrated is called *hill-climbing* or the *method of steepest descent*. A man trying to climb a hill in a fog, if he went always up the slope the steepest way, would be bound to arrive at the top—some sort of top, anyway. If C_T is the altitude above the x, y plane, then

$$\frac{\mathrm{d}y}{\mathrm{d}x} = \frac{\partial C_T/\partial y}{\partial C_T/\partial x}$$

is the compass bearing of the direction of steepest ascent (Figure 3.7b).

A point x_1, y_1 is chosen and the direction of steepest ascent found. To find the minimum, a further point x_2, y_2 is taken a suitable distance from x_1, y_1 in the reverse direction, that of steepest descent, and the process is repeated. It sounds easy enough, but in practice there are often grave difficulties and many variations on the basic procedure have been devised.

With more than two independent variables the principles of finding minima and maxima are not greatly altered but the practical difficulties increase rapidly.

When there are only two independent variables and the problem is an ordinary well-behaved engineering one, a brutal process of calculating C_T for

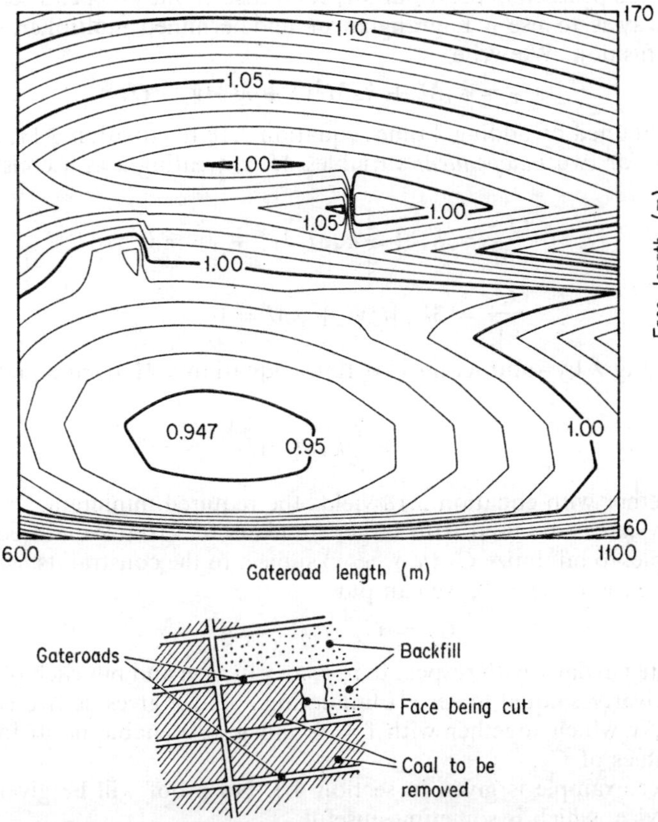

Figure 3.8 Optimization of parameters in coal mining. Here the cost per ton of coal won is plotted against the length of face being cut and the length of the gate roads (see small figure) in retreat mining with reusable props. The calculations have been made and the curves drawn by a computer. (By kind permission of the National Coal Board)

various sets of values of x and y is often justified. This is what has been done in Figures 3.1 and 3.8. With a computer the cost may not be too high and the information contained in such plots is a great deal more useful than a mere knowledge of the minimum.

(c) With several variables and constraining equations

Consider the problem of finding the minimum of

$$C_T = k_1 M + k_4 M^{\frac{2}{3}} v^3 \tag{3.17}$$

where M and v are variable but subject to the constraining equation

$$Mv - Q = 0 \tag{3.18}$$

We could eliminate either M or v from equation 3.17 by means of equation 3.18, and then proceed in the usual way for a case of one independent variable. Another way is to use a Lagrangian or undetermined multiplier, λ, in the following fashion. We write

$$C_T = k_1 M + k_4 M^{\frac{2}{3}} v^3 + \lambda(Mv - Q) \tag{3.19}$$

which is obtained by adding λ times equation 3.18 to equation 3.17, and then proceed as for two *independent* variables M, v, treating λ as a constant. We find that

$$\frac{\partial C_T}{\partial M} = k_1 + \frac{2}{3} k_4 M^{-\frac{1}{3}} v^3 + \lambda v = 0 \tag{3.20}$$

$$\frac{\partial C_T}{\partial v} = 3k_4 M^{\frac{2}{3}} v^2 + \lambda M = 0 \tag{3.21}$$

Eliminating λ by subtracting v/M times equation 3.21 from equation 3.20 gives

$$k_1 = \frac{7}{3} k_4 M^{-\frac{1}{3}} v^3$$

which together with equation 3.18 yields the required minimum.

This device can be used with larger numbers of variables and equations. For example, to minimize $C_T\,(u, v, w, x)$ subject to the constraints $\mathrm{f}(u, v, w, x) = 0$ and $\mathrm{g}(u, v, w, x) = 0$, we can put

$$C_T = C_T - \lambda \mathrm{f} - \mu \mathrm{g}$$

differentiate partially with respect to u, v, w, x in turn and put each of the four partial derivatives equal to zero. Eliminating λ and μ gives us two equations in u, v, w, x which together with $\mathrm{f} = 0$ and $\mathrm{g} = 0$ enable us to locate the turning values of C_T.

A further example is given in section 4.3. No proof will be given of this elegant device, which is sometimes useful.

(d) With an infinite number of degrees of freedom

A problem of historical importance was this; given two points, A and B, B lower than A, join them with a smooth wire such that a small bead, threaded on it and released from A, reaches B in the shortest possible time. This is a minimization problem in which we have to find, not the values of a finite number of variables u, v, w etc. but a *shape* or *function*, the shape of the wire. This *brachistochrone* (= short time, Greek) problem is an optimization problem with an infinite number of degrees of freedom.

The appropriate analytical tool for this problem is the *calculus of variations* (see for example reference 14), which gives a differential equation for the form of the wire. This kind of optimization rarely arises in engineering problems.

Sometimes a physical idea will lead to an alternative means of solving such problems, as in the next section and the case mentioned in section 4.4.

3.11 The design of rotating discs: an optimum structure

(1) The non-rotating disc

Suppose we need a structure to resist a uniform radial outward load $2\pi P_0$ around the periphery of a circle of radius a, as in Figure 3.9. Stresses are to be

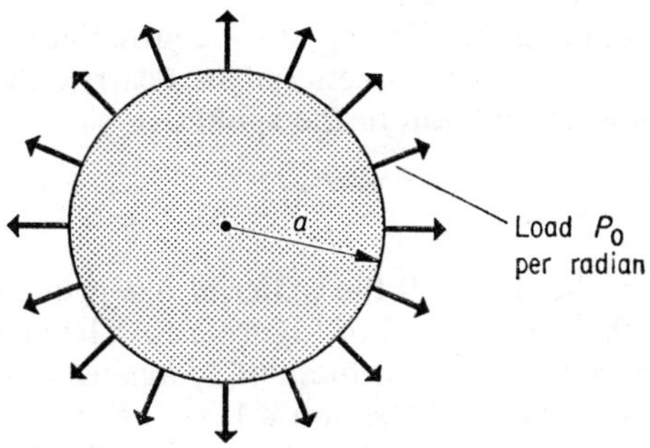

Figure 3.9 Uniformly distributed radial loading

based on a maximum shear stress criterion (Tresca) with an allowable tensile stress f. The allowable combinations of plane principal stresses are shown in Figure 3.10.

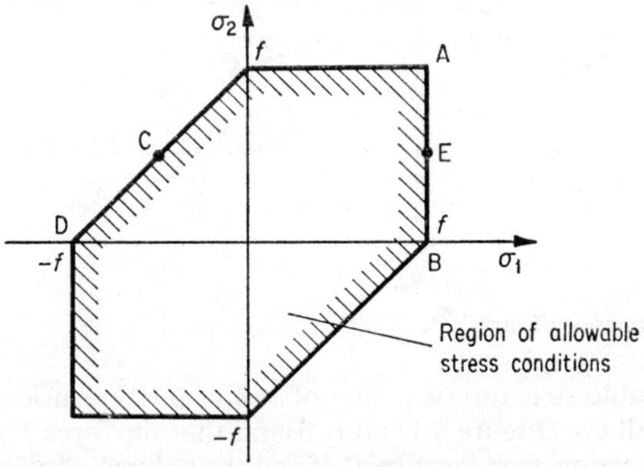

Figure 3.10 Allowable stress, maximum shear stress (Tresca) criterion

Such a load can be sustained by a disc of uniform thickness t, which will be stretched like a drumskin equally in all directions. By equating the load to the inward force at the rim and putting the radial stress equal to f,

$$2\pi P_0 = 2\pi atf$$

$$\text{and } t = \frac{P_0}{af}$$

There is a *load diffusion problem* in getting the load into the disc, requiring a rim, but we shall ignore this.

This simple disc, is, in fact, a structure of minimum weight. The volume of material in it is

$$\frac{\pi P_0 a}{f}$$

The principal stresses at any point in it are $(f, f, 0)$ so the material is all working at point A (Figure 3.10).

Now consider a set of spokes joining at the centre (let us ignore how this is to be done). Their combined cross-sectional area carries the load $2\pi P_0$ and so is $2\pi P_0/f$. The volume of material in the spokes is

$$\frac{2\pi P_0 a}{f}$$

i.e. just twice that in the disc. All the material has principal stresses in it of $(f, 0, 0)$, i.e. it is working at point B of Figure 3.10. The material of the disc is doing 'holding together' per unit volume of $2f$ against f for the spokes. The '*structural task*' to be performed (section 8.1) is $2\pi P_0 a$, and dividing it by the appropriate 'holding together per unit volume' gives the respective volumes of material required.

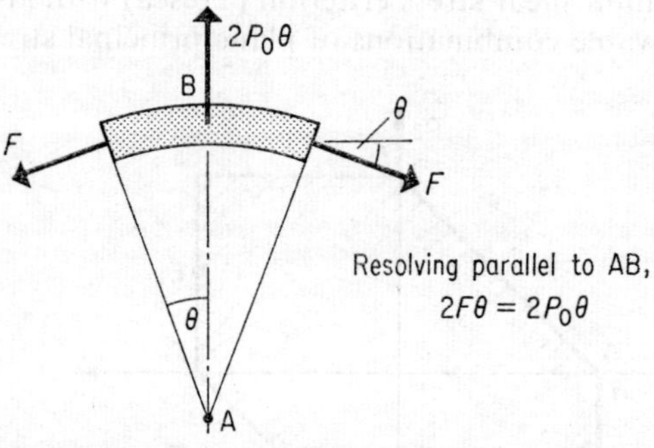

Figure 3.11 Element of ring

A third possible structure is a ring of radius a. By considering the equilibrium of a small arc (Figure 3.11) it is found that the force F in the ring is P_0 and hence the section area must be P_0/f and the volume of material

$$\frac{2\pi P_0 a}{f}$$

a result we could have predicted from the fact that the material is stressed in one direction only.

Of all the permissible working points in Figure 3.10, A has the maximum 'holding together' per unit volume, and all the material in the disc is working at it: the structure is therefore one of minimum weight.

(2) The effect of centrifugal forces

The ideal form of structure to sustain the load system of Figure 3.9 when there are no loads due to the material itself has been shown to be the flat disc of thickness

$$t = \frac{P_0}{af}$$

The commonest occurrence of large-load systems of this sort, however, is in rotating discs, and there the centrifugal forces on the disc material make the problem more complicated. It can however be solved very readily by a small extension of the original treatment.

If the angular velocity of the disc is ω and the density of the material is ρ, then the centrifugal load per unit volume on material at radius r is $\rho\omega^2 r$. The volume of disc lying between radii r and $r + dr$ is $2\pi rt dr$ (where t, the disc thickness is no longer constant but an unknown function of r).

The total centrifugal load on this elementary ring of material is $2\pi\rho\omega^2 r^2 t dr$. Now this load is uniformly distributed round a circle of radius r, so that it is the same kind of load system as that of Figure 3.9 and we have already seen that the ideal structure to carry such a load system is a disc of thickness

$$\frac{\rho\omega^2 r^2 t dr}{rf} = \frac{\rho\omega^2 rt dr}{f}$$

Thus at the inside edge of the elementary ring the disc needs to be just this amount thicker than at the outer edge. In other words, the *decrease* in thickness from r to $r + dr$ is

$$-dt = \frac{\rho\omega^2 rt dr}{f} \tag{3.22}$$

and this same relation must hold over any other elementary ring. Writing equation 3.22 in the form

$$-\frac{1}{t}\frac{dt}{dr} = \frac{\rho\omega^2 r}{f}$$

and integrating

$$-\ln t = \frac{\rho\omega^2 r^2}{2f} + \text{a constant.} \tag{3.23}$$

Putting in the end condition $t = P_0/af$ at $r = a$,

$$t = \frac{P_0}{af}\exp\left\{\frac{\rho\omega^2}{2f}(a^2 - r^2)\right\} \tag{3.24}$$

giving the kind of profile shown in Figure 3.12. This design also is of minimum weight.

The formal statement of the problem we have solved is: given the loading system of Figure 3.9, rotating with angular velocity ω, find the structure of minimum weight that will support it, subject to the allowable stress condition of Figure 3.10. This is a problem of determining an optimum shape, what we called in section 3.10 a brachistochrone type of problem, and we have solved it without using any advanced mathematics. We have been able to do this because of a special property of the problem; there exists a measure ($2\pi P_0 a$) of the structural task to be performed. If σ_1, σ_2 are the principal stresses in the

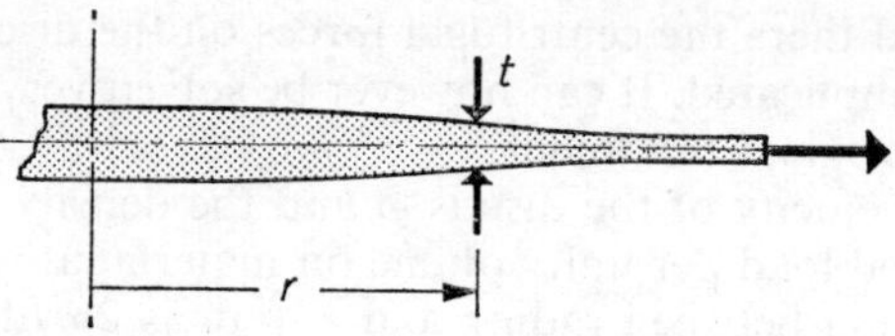

Figure 3.12 Uniformly stressed disc

disc (regarded as thin, so that the axial stress is zero) then in the case of the non-rotating disc

$$\int(\sigma_1 + \sigma_2)\mathrm{d}v = 2\pi P_0 a$$

where dv is an element of volume and the integral is taken over the whole volume of the structure. Thus we have only to make ($\sigma_1 + \sigma_2$) take everywhere the maximum permissible value by making the whole of the material work at point A of Figure 3.10 and we shall have a minimum-volume design. This sort of rule for obtaining an optimum design we shall call an autonomous local optimizing principle, or A.L.O.P. for short—autonomous and local because it can be applied at a point regardless of anything that happens elsewhere. Such principles are of little practical use, but may be powerful aids to developing insight.

It is not always possible to satisfy the A.L.O.P. at all points, as in the problem of the next section.

3.12 Hub design

To complete this brief study of an optimal structure, the rotating disc, we shall consider the common case where a central hole of radius b must be provided. For simplicity the centrifugal forces will be regarded as negligible so that the ideal structure (in the absence of a central hole) is the uniform disc of thickness t.

Elastic design

By our earlier arguments we want to keep each point of the original disc in the same stress condition (f, f, 0), and this means it will have to exert a radial force tf per unit length or btf per radian round the edge of the hole, as in

Figure 3.13. Some structure will have to be provided to carry this load: in the unbored disc it was the central portion and this is the ideal structure for the purpose. Of the alternative structures we have studied only the ring is suitable, so this is the first thing to try. (The disc will be assumed to be thin, so that the radial extent of a suitable ring is small.) If the ring is of cross section A_h, then the hoop stress σ_h in it is

$$\sigma_h = \frac{\text{load per radian}}{\text{area}} = \frac{btf}{A_h} \text{ (see section 3.11)} \tag{3.25}$$

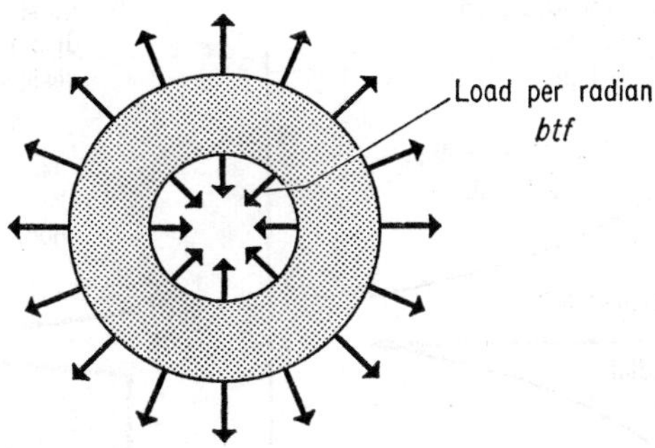

Figure 3.13 Loading of hub of disc

It will not do, however, to put $\sigma_h = f$ if we want the stress to remain everywhere less than f, i.e. if the disc is to be *elastically* designed. There is a relationship between the circumferential strain in the ring (henceforth called the hub) and the stress in it, i.e.

$$\text{circumferential strain in hub} = \frac{\sigma_h}{E} \tag{3.26}$$

When the desired stress condition $(f, f, 0)$ holds in the disc (henceforth called the web), the circumferential strain in it is everywhere

$$\frac{f}{E} - \frac{\nu f}{E} = (1 - \nu)\frac{f}{E}$$

where ν is Poisson's ratio. For the hub to match this the two circumferential strains must be the same, i.e.

$$\frac{\sigma_h}{E} = (1 - \nu)\frac{f}{E} \tag{3.27}$$

and so from equation 3.25

$$A_h = \frac{bt}{1 - \nu} \tag{3.28}$$

If A_h has some other value than this, the circumferential strain in the hub will match that in the web—it must as they are adjacent portions of the same piece of metal—but the stress condition in the web will not be the desired ($f, f, 0$). Figure 3.14 shows the stress distribution in a disc in which the value of A_h is less than that for the elastic design (curves II) and it will be seen that the hoop stress exceeds f throughout. Figure 3.14 also shows the stresses in an *elastically* designed disc (i.e. with hub cross-section given by equation 3.28). The stress condition in the hub of this disc ($\{1-\nu\}f, 0, 0$) seems uneconomically low, and it may be suggested that if the reinforcing (hub) material were

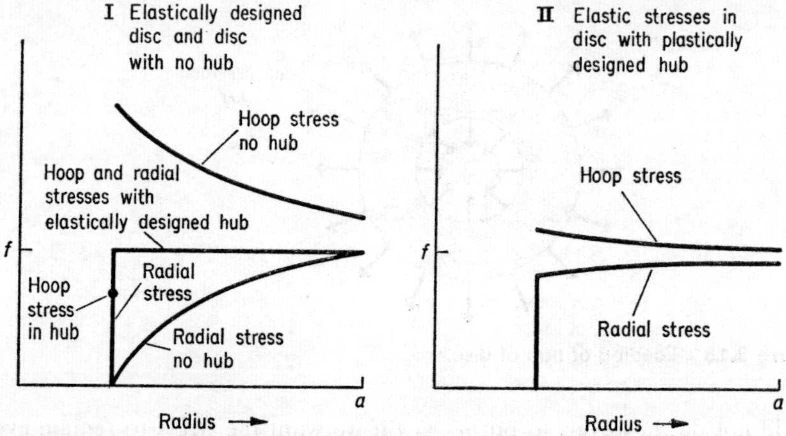

Figure 3.14 Elastic stresses in rotating discs

extended to larger radii it might be more highly stressed and result in a lighter design, still without exceeding the allowable stress f. It is not so.

The elastically designed disc, with hub cross-section given by $A_h = bt/1-\nu$, is a minimum weight structure design to the allowable elastic stress condition of Figure 3.10. In a *plastically designed* disc, we keep the *plastic* stresses within the same boundary, but the strain is assumed to be plastic not elastic, so that the elastic compatibility equation 3.27 need not be met. We put $\sigma_h = f$ in equation 3.25 (which is one of equilibrium), so that $A_h = bt$, giving the elastic stresses we have already seen in Figure 3.14 (curves II).

Plastic design

Suppose the peripheral load on the disc is due to centrifugal forces on blading. Imagine the speed increased from ω to $k^{\frac{1}{2}}\omega$, so that all the loads, being proportional to the square of the speed, are increased by a factor k where

$$k = \frac{\text{u.t.s. of material}}{f}$$

Then $k^{\frac{1}{2}}\omega$ is about the speed at which the elastically designed disc will burst and k is an ultimate factor at speed ω.

Now when overspeeded a disc may burst in either of the patterns shown in Figure 3.15(a) and (b), and the circumferential failure in pattern (a) may have any radius r within the web. The idea of plastic design is to make all these failures equally likely, an old principle exemplified by the one-horse shay. We assume that the material is ductile enough to ensure that before failure takes place the stress on the failing section will become practically uniform by plastic flow. The design principle then reduces to making the average normal stress on any section such as ABCDE or FGHI the same. In this case the average normal stress must be f at normal speed if we require an ultimate factor of k (average here means average with respect to the projected area on the diametral plane AE or FI; this assumes that there are no shear stresses on

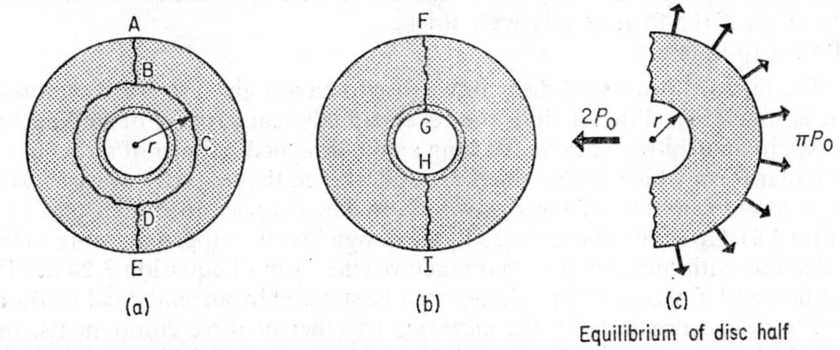

Figure 3.15 Plastic failure of discs

the section, which follows from symmetry). This is a simple equilibrium condition, and in the case where centrifugal forces on the material of the disc are negligible it reduces to (Figure 3.15(c))

$$2P_0 = \text{projected area of section} \times f,$$

i.e. the projected area of section is constant at $2P_0/f$. Now the projected area of any section such as ABCDE is constant at $2at$ if t, the thickness, is constant, so the web is uniform and

$$t = \frac{P_0}{fa}$$

The projected area of section FG–HI is

$$2t(a - b) + 2A_h = \frac{2P_0}{f}$$

so that $A_h = bt$ and the plastic design is complete.

If the centrifugal forces on the material must be taken into account, the thickness of the web can be shown to be given by equation 3.24, just as in the elastic design, and

$$A_h = bt\frac{f}{f - \rho\omega^2 b^2} \tag{3.29}$$

The term $f - \rho\omega^2 b^2$ in equation 3.29 represents a valuable concept we shall not discuss at length. It is the 'effective hoop strength' obtained by subtracting from f the stress needed to support the centrifugal forces on the material itself. An alternative method of deriving equation 3.24 can be based on this measure of usefulness.

The methods described enable the plastic and elastic design of discs to be done very simply and with adequate accuracy provided the thickness is small compared with the diameter. Slight modifications are needed in practice to deal with the rim, which is usually thicker than the web, to enable the load diffusion from the blades to be achieved with joints of practicable efficiency, and the finite radial extent of the hub. The design of discs is more difficult if the material properties vary with the radius or large thermal stresses are present, or if the form is relatively thick.

Points to note are:

(1) The forward process of designing a disc to have a given stress distribution in it is much less difficult than the reverse analytical process of finding the stresses in an arbitrary design, so long as the designed pattern of stress has a particular form which is also the most desirable in the majority of cases. It is the purposeful nature of the disc which makes it easier to design (see also section 4.4). However, the last step in the design involves approximating to the ideal shape with one cheap to manufacture (the form of equation 3.24 hardly qualifies) and if necessary this shape must be stressed by an analytical method.
(2) The design proceeds by the *matching* together of three components, the rim (which we have omitted), the web (with which we have dealt fully), and the hub (with which we have dealt rather inadequately). In this respect the problem of disc design is rather like one of systems engineering.

3.13 Importance of optimization in design

Large-scale formal optimizations are growing in importance in engineering design, e.g. in aerospace structures. Small optimizations of detail could probably be used much more with great advantage.

Simple studies of the kind made in this chapter are not accurate enough for problems where large sums of money are involved, but will do for small jobs (see section 10.2) and are often valuable for developing insight. Simple studies also provide useful guidance in a common sort of large problem where there are many degrees of freedom and an accurate cost function is too elaborate to be tractable by formal methods.

Questions

Q.3.1. (2) Using the same assumptions as section 3.3 (e.g. drag proportional to v^2) prove the result given there for a ship optimized for a fixed number of tonne-kilometres per year, i.e. the ratio of a to b type costs is 7/3.

Q.3.2. (3) For a ship of given tonnage, on the assumptions of section 3.3, what is the effect on optimum speed of allowing for a turn-round time t per voyage of length s? Will it be higher or lower than for the case $t = 0$?

Q.3.3. (4) For a given number of tonne-kilometres per year, will the optimum speed of a ship be higher or lower if a turn-round time t per voyage of length s is allowed for, compared with the case $t = 0$? Answer for the two cases (a) t constant and (b) t proportionate to tonnage M, and for both state the nature of the critical term in the costs which is responsible for the result in question. (This question and Q.3.2 can be answered by reasoning alone, but a very high standard of insight is required, or else hindsight after doing the calculations).

Q.3.4. (2) Electrical power P is to be supplied at a d.c. voltage of about V over a distance s, a return cable being used. The conductors are of specific resistivity ρ and cost c_1 per unit volume year. Power costs are c_2 per unit year. Find the minimum total cost, assuming this to be small compared with that of the power transmitted.

Q.3.5. (3) A uniform beam AB, pinned at each end and of length L, carries a uniformly distributed load w per unit length and a concentrated load wLy at a point C such that AC $= Lx$. Plot on an x, y graph the zone in which the maximum bending moment occurs at C.

Q.3.6. (2) Assuming the validity of equation 3.8, what percentage increase in capital cost of a heat exchanger is necessary to reduce the temperature difference by 1 per cent, without changing the pressure drop?

Q.3.7. (4) Two depots B and C are each to be supplied with oil at the same steady rate from a third depot A; A, B and C form an equilateral triangle. A pipeline is to be laid from A to some point D on the perpendicular bisector of BC. From D two smaller pipes are to be led to B and C. Find the value of the angle BDC which minimizes the sum of pumping and pipe costs, on the following assumptions:

Pumping costs are proportional to pressure loss, which is governed by the pipe friction formula with a constant friction factor, so that pressure drop per unit length is proportional to velocity squared and inversely proportional to pipe diameter. Pipe costs per unit length are proportional to diameter (hint: first find the relationship between flow rate and optimum diameter).

Q.3.8. (2) An axial piston swashplate type hydraulic pump (see Figure 3.16) has a fixed swash angle of 20° and n cylinders of diameter d. To accommodate

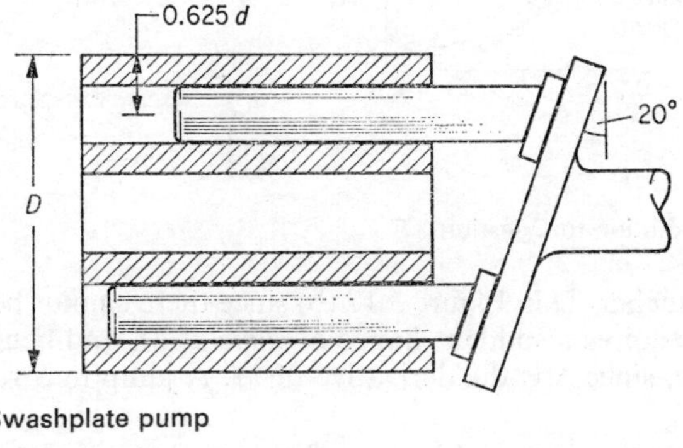

Figure 3.16 Swashplate pump

the slipper pads, the centre distance between the cylinders must be a minimum of 1.25*d* and for the same reason the overall diameter *D* of the cylinder block must exceed the p.c.d. of the cylinders by 1.25*d*. Find the number of cylinders which gives the maximum swept volume per revolution for a given value of *D* and the percentage decrease involved in using (a) one more cylinder and (b) one less cylinder.

Answers and notes

A.3.1. The appropriate form of C_T is given by equation 3.17, where *M* is the tonnage.

A.3.2. Lower: this rather surprising result is the effect of the capital cost of the engines. Remember that the *b* term of section 3.3 involves costs of engines and fuel, and fuel is not burnt when not under way but engine capital charges continue.

A.3.3. (a) lower due to engine capital cost term, higher due to fuel cost term: on balance likely to be nearly unchanged; (b) higher: since the turn-round time is proportional to tonnage, and the total tonnes per year is constant, a fixed fraction of the year is lost. Capital terms in both *a*- and *b*-type costs will be increased accordingly, but not the fuel part of *b*-type costs. Costs for *a* will be increased relatively more, so that the optimum speed rises, as with the L.N.G. tanker.

A.3.4. $4s\sqrt{(c_1c_2\rho)}\,\dfrac{P}{V}$

A.3.5. See Figure 3.17. Inset (a) shows the loading and bending moment (*M*) diagram for a case where the maximum *M* occurs at C, and (b) for a case where it does not. When the maximum does not occur at C it occurs at a

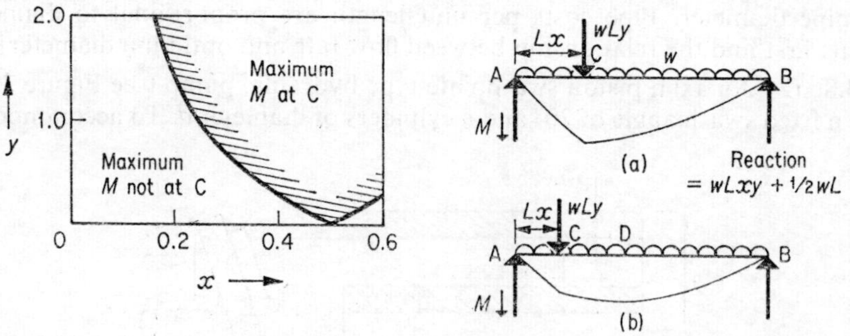

Figure 3.17 Solution to Question 3.5

turning value such as D in Figure 3.17 (b) since there cannot be a cusp except at C (A cusp requires a sudden change of *slope* of *M*, and hence a *jump* in *S*, the shear force, since *S* is the derivative of *M*. A jump in *S* requires a point load).

The regions of x, y in which the maximum M is at the cusp at C and those in which it is at a turning value elsewhere are separated by a curve where the cusp and the turning value coincide. A turning value of M implies a zero of S, and so on the boundary curve $S = 0$ just towards the centre from C. From Figure 3.17 (a),

$$S(x+) = wLxy + \tfrac{1}{2}wL - wL(1 - x) = 0$$

on the boundary curve, giving

$$xy + x = \tfrac{1}{2}$$

A.3.6. 1.4 per cent. From equation 3.8, for a change x, y, not to alter π, the pressure drop term $x^{-1.8}y = 1$, so that if p is the small percentage increase in y, that in x must be $p/1.8 = 0.55p$.

A.3.7. 90°. The optimum diameter is proportional to the square root of the flow rate, so that the branches are $1/\sqrt{2}$ times the diameter of the main and $1/\sqrt{2}$ times the cost per unit length. The pumping costs are 0.2 times the pipe costs. The total cost is proportional to

$$\text{AD} + 1/\sqrt{2}\,\text{DB} + 1/\sqrt{2}\,\text{DC} = \text{AD} + \sqrt{2}\text{DC},$$

which is readily minimized. Indeed, by sketching two close trial positions of D it can be seen at once where the minimum occurs.

A.3.8. Seven cylinders (a) 0.9 per cent (b) 0.6 per cent.

4
Insight

4.1 Introduction

Perhaps the most important single prerequisite for good solutions to design problems is insight (see section 1.4). From numerous instances [2] it may be inferred that insight frequently develops by large steps—it 'dawns' or 'comes in a flash'—but the steps are separated by laborious stretches of mental spadework.

By the use of aids we can hope to do two things. Firstly, we may be able to reduce the amount of spadework. But sometimes even the spadework does not suggest itself, and then we may be glad of a line on which to work, laborious or not.

This chapter discusses some approaches and devices which may be effective either in developing insight more speedily, or in developing it at all in more difficult cases (e.g. section 4.4).

4.2 Rough calculations

A good general principle is to quantify whenever possible. It is easy to waste much time on qualitative studies of matters which might be clarified by a few lines of arithmetic. Designers should cultivate the habit of making very rough quick calculations whenever the need arises, and they should not be discouraged too easily by superficial difficulties from doing so.

Chesterton, I think it was, who said that if a thing is worth doing it is worth doing badly, an aphorism which is very true of engineering calculations. In the early stages of a design there is often little need of accuracy, and it is wasteful to make lengthy computations or to have recourse to a computer when a few minutes with a slide rule will give an adequate answer. Sometimes the order of magnitude alone is enough to show that an effect can be safely ignored or that a line of thought must be abandoned, as in the case of the ice-covered methane tanker (section 1.4). An accurate calculation done at an early stage will usually have to be repeated towards the end with different data, and is only justified if it is vitally concerned with the feasibility of the scheme. Even then, it may often be preceded with advantage by several 'rounds' of quicker, rougher calculations which may help in 'homing in' on a suitable set of trial parameters. Even when conversational mode on-line computing becomes generally available, it seems unlikely that all the needs of the designer will be met, and some skill in rough calculation will still be useful. Some examples of such work will be given.

(1) Diaphragm clamp

It is required to clamp a slide of a machine tool hydraulically by means of a circular diaphragm as shown in Figure 4.1. Suitable values for dimensions t

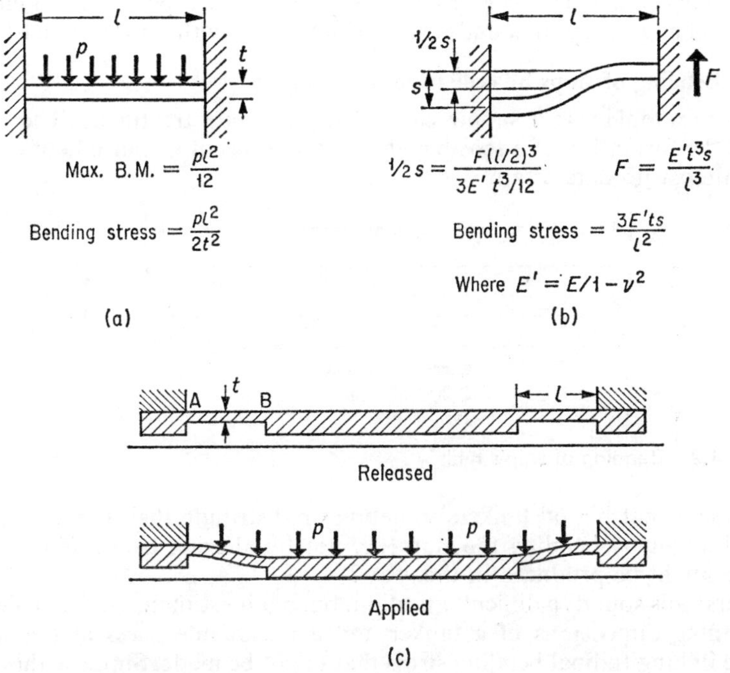

Figure 4.1 Diaphragm clamp

and l are to be determined, given the operating gauge pressure p, the radius r, the maximum travel s, and the allowable stress f.

To design the diaphragm using the stress analysis for a circular plate would be tedious, but for practical purposes we can regard the thinned zone as a beam, subject to two forms of loading, that due to the imposed load, the pressure p (Figure 4.1(a)), and that due to the imposed deflection s (Figure 4.1(b)). The only account we need take of the great width of the beam in the circumferential direction is that circumferential strains will be suppressed (i.e. the beam approximates to a condition of plane strain rather than plane stress), and this means that we must substitute

$$E' = \frac{E}{1 - \nu^2}$$

where ν is Poisson's ratio, for E in the usual beam deflection formulae.

It is easily found that the maximum tensile stress, σ_M, which occurs on the oil side at A, is given approximately by

$$M = \frac{pl^2}{2t^2} + \frac{3E'ts}{l^2} \tag{4.1}$$

and if r is a few times greater than l, the accuracy will be adequate for all useful purposes. It is much easier to see from this expression the effect of various changes, say, in s and t, than it would be with a more exact form.

(2) Bending of ships by redistribution of cargo

The last example was a simple case of approximate treatment. The next is one of the formation of a sensible idea of the order of magnitude of an effect from almost no data at all.

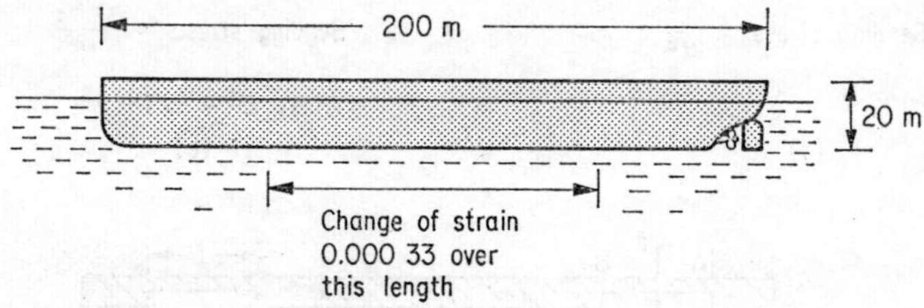

Figure 4.2 Bending of ship's hull

It is said that big oil tankers sometimes redistribute their cargo to reduce their draught by bending the ship (i.e. not by altering trim). What sort of change might be produced in this way?

At first this sounds a difficult question, but a fair estimate can be made from the leading dimensions of a tanker and a reasonable guess at the sort of change in longitudinal bending stress that might be made. Suppose this might average as much as 60 N mm^{-2} over the middle half of the ship's length—the change in the ends is not very significant. Regard the ship as a beam (Figure 4.2), and apply Euler's theory of bending summarized in the well-known equation

$$\frac{M}{I} = \frac{f}{y} = \frac{E}{R}$$

where M is bending moment, I is the second moment of area of section, f is longitudinal stress, y is distance from the principal axis of the cross section and E is Young's modulus. Then the radius of curvature R induced in the ship is given by

$$R = \frac{Ey}{f} = 3300y$$

taking $E = 200$ kN mm^{-2} for steel.

Now the depth of the hull of such a ship might be about 20 m, and if we take y as half this

$$R = 33\,000 \text{ m}$$

Thus if the keel had a radius of curvature, or rocker, of about this much, it might be feasible to remove it by redistributing cargo, without inducing excessive stresses. The difference in draught between ends and middle, h in Figure 4.3, for such a curvature is roughly (Pythagoras' and binomial theorems)

$$\frac{1}{2R}\left(\frac{l}{2}\right)^2$$

which if l is 200 m and R is 33 000 m gives 0.15 m. However, this is not the decrease in draught, since the displacement must remain constant: as the centre sections rise, the end sections must sink. If the waterline plan of the

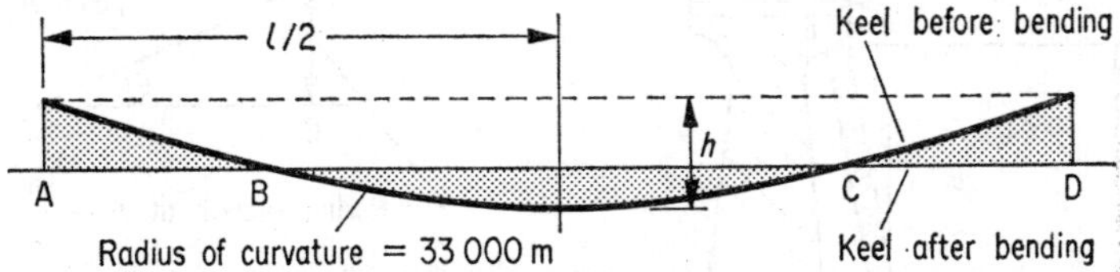

Figure 4.3 Bending of ship's hull: effect on draught

ship were a rectangle, the shaded area shown in Figure 4.3 would have equal amounts above and below the line representing the straightened keel: then the volume of ship raised from the water between B and C would be equal to the volume lowered in between A and B and between C and D, the displacement remaining unchanged. Since the mean height of a parabolic segment is two-thirds the maximum height, the reduction in draught is $h/3$.

However, since the waterline beam at the ends falls away rapidly, the shaded areas above the straightened keel must be significantly greater than those below if they are to represent an equal volume; the saving in draught will therefore be considerably less, say, about 0.25 h. If the original curvature is concave downwards (hogged), the reduction will be about 0.75 h, or, say, 0.11 m.

In taking the curvature $1/R$ to extend throughout the length, the value of h is overestimated; if the curvature is restricted to the middle half, the value will drop to three-quarters. On the other hand, the neutral axis of the ship in bending will be nearer the bottom than the deck, so that y should be taken as less than half the moulded depth. All in all, the reduction in draught of a hogged vessel might be expected to reduce by about 0.1 m, in which estimate the chief room for error lies in the mean change in stress (60 N mm^{-2}) which was assumed. (It must be realized that this stress cannot be uniform over the middle half, but must reduce considerably towards the ends.)

It is not suggested that this sort of sketchy calculation is any substitute for accurate work, merely that it is valuable use of the minutes or even seconds it takes. Gross errors in elaborate hand and computer calculations can frequently be detected by such means. Besides this they generate insight into the importance of effects and the way they may be expected to vary with various parameters.

The remaining examples are both taken from one narrow field, that of rotational stresses in disc-like structures, a topic which has already been discussed in Chapter 3.

(3) Turbine cooling air shroud

Figure 4.4 shows part of a gas turbine, with a shroud disc 1 which serves to lead cooling air up the forward face of the bladed disc 2. At the point A at

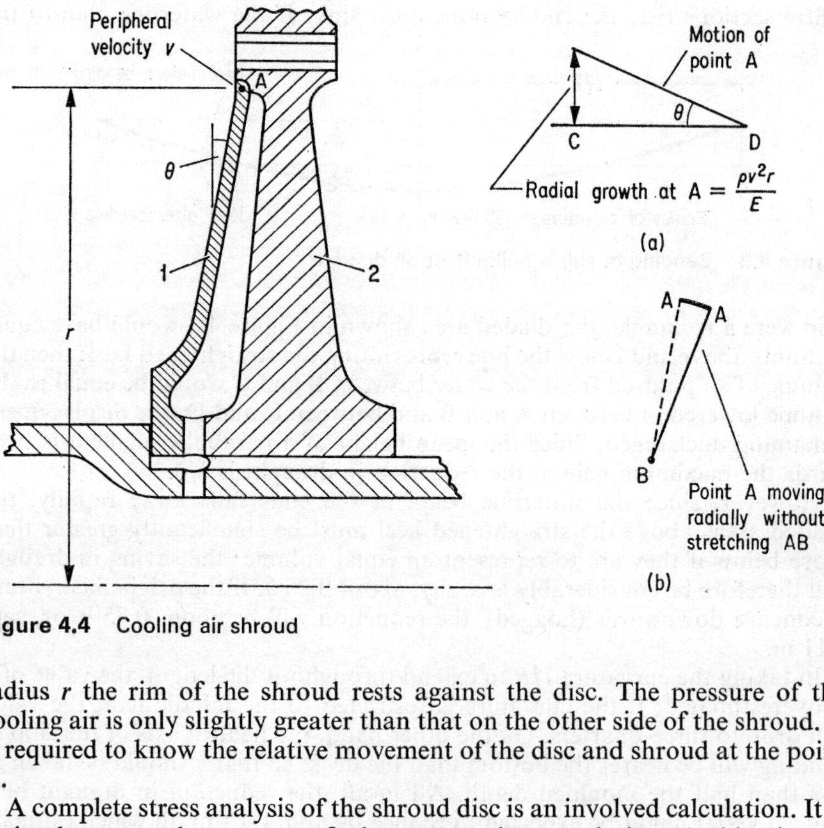

Figure 4.4 Cooling air shroud

radius r the rim of the shroud rests against the disc. The pressure of the cooling air is only slightly greater than that on the other side of the shroud. It is required to know the relative movement of the disc and shroud at the point A.

A complete stress analysis of the shroud disc is an involved calculation. It is a simple matter, however, to find an approximate solution. A thin ring of metal of radius r rotating by itself would have a stress in it of

$$\rho v^2$$

where ρ is the density and v the peripheral velocity (see section 3.12). The strain in it would be

$$\frac{\rho v^2}{E}$$

and the radial growth would be

$$\frac{\rho v^2 r}{E}$$

If such a ring forms the rim of a thin *flat* disc, it cannot move radially outwards this much because the flat disc can only be stretched radially by stretching the web. But a conical disc can be stretched radially much more easily by *bending* of the web as in Figure 4.4(b) and hence much lower forces will suffice. The radial stiffness of the rim of the cone is orders lower than that of the flat disc. Thus we may regard the thin ring at A as virtually unsupported radially, so that the circumferential stress in it and the radial growth are given sufficiently accurately by the expressions above. Also since this relatively free radial growth is achieved by bending the web, it must occur as the radial component of a motion normal to the generators of the cone, as in Figure 4.4(a) and (b). Thus the axial separation of the shroud from the disc is CD in Figure 4.4(a), and

$$\mathrm{CD} = \frac{\rho v^2 r}{E} \cot \theta$$

The writer scored a small success when he first made this calculation in his head, just before an experimental determination was begun, the result of which agreed very closely with his estimate. A second quick estimate is also needed to be sure that it is reasonable to do the first, but this will not be given here.

(4) Thermal stresses in discs

Consider a turbine disc, such as those discussed in section 3.11 or the one in Figure 4.4, with a substantial temperature difference, say, 100°C, between hub and rim. Let us try to estimate, very roughly, the resultant thermal stresses. If the material is steel with a Young's modulus of 200 kN mm^{-2} and a mean coefficient of thermal expansion of 12×10^{-6} per °C over the range concerned, then the thermal strain caused by 100°C temperature change is 1.2×10^{-3}, which could be produced by a tensile stress of 240 N mm^{-2}. The outer, hotter, part of the disc will try to grow in diameter, tending to stretch the inner part to fit it. The resisting pull of the inner part will decrease the expansion of the outer part below the 1.2×10^{-3} strain it would adopt if unrestrained. The mismatching thermal strain of the rim relative to the hub will be accommodated in three ways:

(a) by pulling in the rim, which will be in *compression* circumferentially,
(b) by stretching the web, radially, and
(c) by stretching the hub, which will be in *tension* circumferentially.

If only a narrow rim of the disc is hotter than the rest, as in the temperature distribution I of Figure 4.5, then it is clear that nearly all the mismatch will be taken up in way (a), since the amount the small hot portion will be able to stretch the remainder will be negligible. The rim will thus be subject to a compressive stress of almost 240 N mm^{-2}.

In the case of a temperature distribution such as II in Figure 4.5, we may regard the differential thermal strain as equivalent to an effective stress of 240 N mm^{-2}, of which, say, 130 N mm^{-2} appears as a compressive stress at the outside, 30 N mm^{-2} is accounted for by tensile stress at the inside, and the remainder is accounted for by radial stretching of the web. Because the circumferential length of the hub is less than that of the rim, the actual tensile

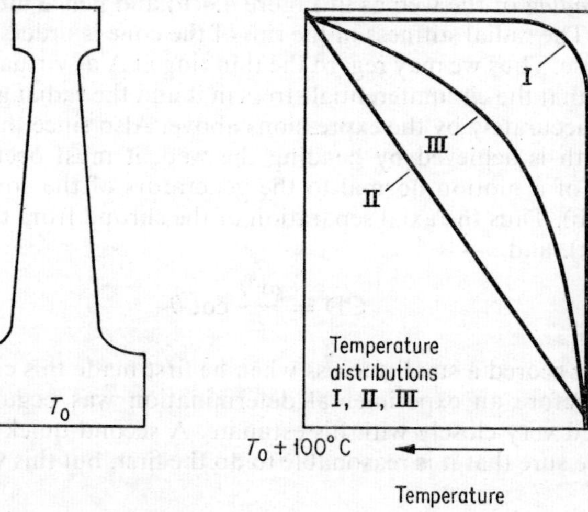

Figure 4.5 Thermal stresses in discs

stress in it will be much higher than 30 N mm^{-2}. The appointment of the effective stress between rim and hub must be done in inverse ratio to the areas of the sections, in order to satisfy conditions of equilibrium. A similar argument determines the allocation to the stretching of the web.

Of course, this kind of argument gives only a rough picture of the sort of stresses to be expected in the disc, but it enables us to develop a feel for the problem, so we can guess very well what will happen to the stresses if the temperature distribution changes, say, to III in Figure 4.5, or we make the rim lighter in section, and so forth. Anyone who doubts the respectability of the 'quick sum' should consider the following passage.

> Let it be assumed that the typical night-bomber is a metal winged craft well bonded throughout, with a span of the order of 25 metres. The wing structure is, to a first approximation, a linear oscillator with a fundamental

wavelength of 50 metres and a low ohmic resistance. Suppose a ground emitting station to be set up with a simple horizon half-wave linear oscillator perpendicular to the line of approach of the aircraft and 18 metres above ground. Then a craft flying at a height of 6 km and at 6 km horizontal distance would be acted on by a resultant field of about 14 millivolts per metre, which would produce in the wing an oscillatory current of about $1\frac{1}{2}$ milliamperes per ampere in sending aerial. The re-radiated or 'reflected' field returned to the vicinity of the sending aerial would be about 20 microvolts per metre per ampere in sending aerial.

This passage occurs, with others like it, in a historic paper on the detection and location of aircraft by radio methods—radar, in fact—presented to the Committee for the Scientific Survey of Air Defence on 27 February 1935.

Another historic 'quick sum' is given by Watt in a letter written to Boulton about the threat to their engine from a turbine. He deduced that the turbine blades would have to move at very high speeds to be efficient and wrote [15]:

> In short without god makes it possible for things to move 1000 ft pr″ it can not do much harm.

This curt dismissal of the steam turbine as a practical possibility is an entirely adequate summary of the situation then and would merely have lost force by elaboration. It does not even matter that it would have been enough to be able to make things to move 400 ft pr″.

4.3 Optimization of compressor shaft diameter

The problem here treated will be taken as an example of a rough calculation though it is also an optimization (Chapter 3) and a differential study (section 3.6). It was seen in section 2.3 that the shaft of an axial flow compressor rotor of shaft-and-disc construction has a structural function, which demands a section of given bending stiffness per unit length EI, or, if the material is given, a fixed second moment of area of cross-section I.

If the shaft is tubular with inner and outer radii r_I and r_0, respectively (Figure 4.6),

$$I = \frac{\pi}{4}(r_0^4 - r_I^4) \tag{4.2}$$

If r_0 is increased while keeping I fixed, the cross-sectional area

$$A = \pi(r_0^2 - r_I^2) \tag{4.3}$$

decreases and the shaft becomes lighter. However, the holes in the discs become larger and the hubs heavier. In an aircraft, turbine weight is important, so that we seek the value of r_0 that gives the minimum combined weight of shaft and hubs. In what follows it will be assumed that shaft and discs are of the same density, so that the problem becomes one of minimizing volume. Consider the length of shaft p occupied by one disc of web thickness t.

Now it was shown in section 3.12 that in an elastically designed disc the cross-sectional area of the hub reinforcement A_h (shown shaded in Figure 4.6) is given by

$$A_h = \frac{r_0 t}{1 - \nu} \tag{4.4}$$

where ν is Poisson's ratio. The volume of the hub reinforcement V_h is $2\pi r_h A_h$, where r_h is the mean radius of the reinforcement (see Figure 4.6) and will be fractionally greater than r_0, say, $1.05r_0$, so that, if ν is 0.3,

$$V_h = 2\pi \times 1.05 r_0 \times \frac{r_0 t}{0.7} = 3\pi r_0^2 t \tag{4.5}$$

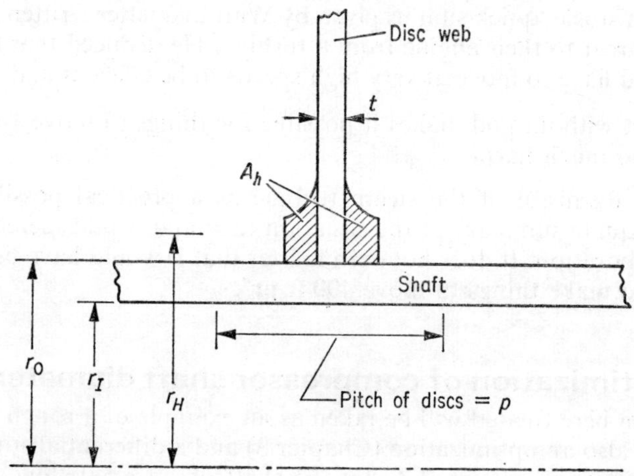

Figure 4.6

But the volume of web cut out in making the hole is $\pi r_0^2 t$, so that the net volume penalty over a disc with no central hole is $2\pi r_0^2 t$. We might say that the weight of a hole is twice the weight of material cut out.

The additional volume associated with the hub is $2\pi r_0^2 t$ and the disc occupies an axial length p (Figure 4.6). The additional hub volume per unit length of shaft

$$\frac{2\pi r_0^2 t}{p}$$

is a kind of effective or mean cross-sectional area which may be added to the shaft cross-sectional area. (The thickness t of the webs and the pitch p of the discs will vary through the compressor, but an average t and an average p will involve no error.) We require to minimize this combined cross-sectional area A_T subject to a constant value of I. This we do by minimizing $A_T + \lambda I$, subject to no restriction, by varying r_0 and r_1 (see section 3.10) where

$$A_T = \frac{2\pi r_0^2 t}{p} + \pi(r_0^2 - r_I^2) \tag{4.6}$$

We can just as well minimize

$$U = \frac{4}{\pi}(A_T + \lambda I) = \frac{8r_0^2 t}{p} + 4(r_0^2 - r_I^2) + \lambda(r_0^4 - r_I^4) \tag{4.7}$$

and we can differentiate with respect to r_0^2 and r_I^2 rather than r_0 and r_I

$$\left.\begin{aligned} \frac{\partial U}{\partial(r_0^2)} &= \frac{8t}{p} + 4 + 2\lambda\, r_0^2 = 0 \\ \frac{\partial U}{\partial(r_I^2)} &= -4 - 2\lambda r_I^2 = 0 \end{aligned}\right\} \text{at the minimum of } U \tag{4.8}$$

Eliminating λ, $\dfrac{r_0^2}{r_I^2} = \dfrac{2t}{p} + 1$ and we find

$$r_I^4 = \frac{\frac{4}{\pi} I \left(\frac{p}{t}\right)^2}{4\left(1 + \frac{p}{t}\right)}, \quad r_0^4 = \frac{4}{\pi} I + r_I^4$$

These results are plotted in non-dimensional form (such that $r_0 = 1$ for a solid shaft) in Figure 4.7. As might be expected, the highest values of p/t (i.e. thin discs like compressor discs) give thin-walled optimum shafts, while thick discs give thick-walled shafts. The argument has been simplified by ignoring

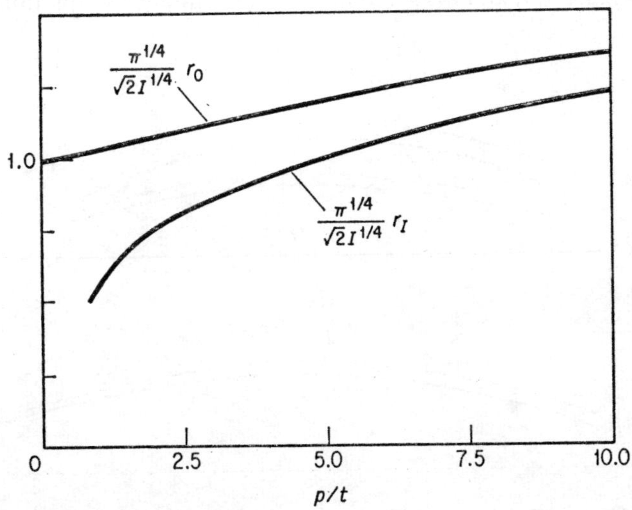

Figure 4.7 Variation of inner and outer radii of minimum weight shaft with pitch/web-thickness (p/t) ratio

the influence the variation of the mass of the rotor has on the value of I required, an effect which causes only slight complication. Other slight modifications are needed to deal with the effect of splines or other details which come between the shaft and the hub, and make r_0 different from the disc bore radius.

An alternative to this procedure, which the writer has known to be used, is to do several almost complete designs of compressor rotor with different shaft diameters, and calculate their weights. This is very expensive, takes much longer and is also much more likely to yield a misleading result. The designers working on the various schemes and the stressmen are quite likely to introduce errors, through small differences which are not associated with the changes in shaft size, through rounding small dimensions to different convenient figures, through differences in taking scaled measurements, or through insufficiently standardized design or stressing methods for the discs etc. The differential study (section 3.6) we have made is cheap, quick, and reliable, because it deals only with the change we are interested in.

4.4 The optimum virtual shaft: a digression

Before leaving this example, let us look at a more advanced case. So far we have regarded I, r_0 and r_I as constant throughout the length of the shaft. But the requisite stiffness could be achieved with less material by making I greater at the mid-length of the shaft and less towards the end. While such a variable cross-section scheme would be expensive to manufacture in a discs-and-shaft construction, it is eminently feasible in some virtual shaft designs. In such an arrangement r_0 and r_I are functions of x, the distance along the shaft (see Figure 4.8) and I is given in terms of r_0 and r_I by equation 4.2, and is

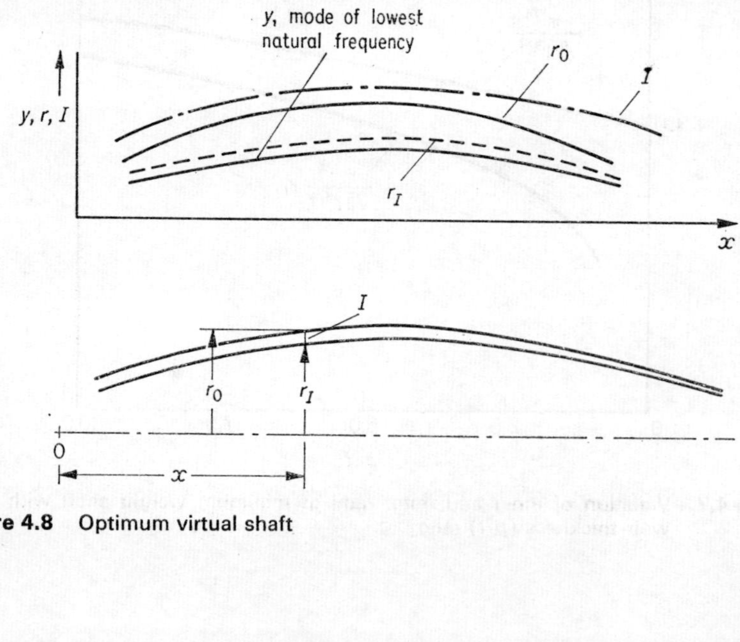

Figure 4.8 Optimum virtual shaft

also a function of x. We now require to choose these functions to give a rotor of minimum weight, subject to the condition that the lowest natural frequency of transverse vibrations of the shaft shall have a prescribed minimum value (see section 2.3). This is a brachistochrone problem (section 3.10) subject to a very complicated constraint. It is too advanced for this book, so mention will only be made of the salient features, which are particularly interesting.

Once we know the value I at any section, we can find the values of r_0 and r_I which will give this I for a minimum weight penalty by the method already described. We specify the shape or mode in which the shaft vibrates, and from this and the required minimum natural frequency decide the value of I corresponding to each value of x. Finally, the mode of vibration is determined from the 'shape' of r_I as a function of x. From a guessed shape of r_I (see Figure 4.8) we determine a corresponding mode y of vibration and hence a shape of I. The actual values of I are then fixed to give the required natural frequency, and r_0 and r_I determined from I by the method already described. This gives in general a different shape of r_I, so the process is repeated until it converges, i.e. until it keeps on giving the same shape of r_I every time.

There are two things to note especially. Firstly, we decide on the shape (mode) and frequency of vibration we want, and hence the required stiffnesses. This is not only a *design*, as against an *analytical*, process; it is much easier than finding the mode and frequency given the stiffnesses. In just the same way we saw in section 3.11 that it is much easier to design a disc to have a given (purposeful) stress distribution than to find the stresses in a given (purposeless) design, the typical analytical problem.

Secondly, the reason for the final design being one of minimum weight lies in the method of deriving the vibratory mode y from r_I. We choose y so that the maximum value of the strain energy of vibration of the material at r_I exceeds the maximum value of its kinetic energy of vibration by the same amount at all x. This is an *autonomous local optimizing principle* (A.L.O.P.—see section 3.11) and automatically produces a rotor of minimum weight.

Unlike the simple method for a uniform shaft, the writer does not know of this design process being used for an existing engine; these few notes are included as an example of the automatic design of an optimum structure.

4.5 Useful measures and concepts

Another good habit for the designer is the use of known convenient measures which help to characterize design elements, and the invention of new ones *ad hoc* to help him in particular problems. Examples are joint efficiency, solidity, aspect ratio, water equivalent, block coefficient—all these are handy measures which help the mind familiar with them to visualize or assess the object or the situation described.

Such terms are so frequently used by all engineers that it may seem unnecessary to emphasize the point. But sometimes the wrong measure is used, or no measure at all where one would be helpful. Sometimes no accepted measure exists, and then the designer should devise his own. We have already had occasion to use such an *ad hoc* measure in the suspension-bridge example of section 3.4, where the cost of the cables involved the multiplier c_c/f_c where c_c is the cost of the material per unit volume and f_c is the working stress. Then the ratio c_c/f_c may be regarded as the cost per unit of strength, and it is a useful concept in the economical design of structures [16].

A common sort of useful measure is a figure of merit for materials for use in a given role. Consider helical compression springs—the wire in them is loaded in torsion, and the maximum stresses occur in the surface, which is in a state of pure shear, i.e. the principal stresses are $(-\sigma, \sigma, 0)$. If these stresses

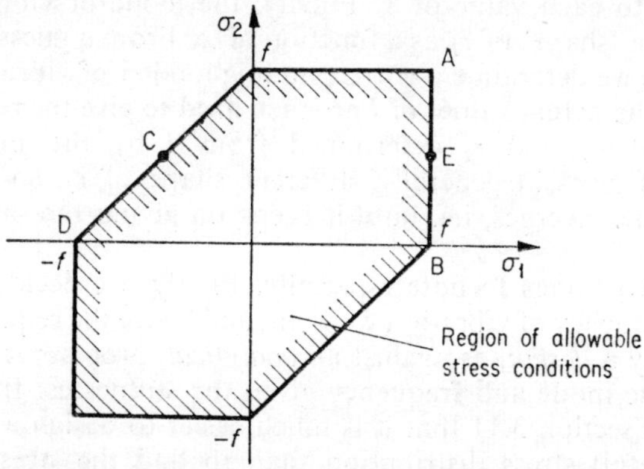

Figure 4.9 Allowable stress, maximum shear stress (Tresca) criterion

are to be the maximum allowable according to some such basis as maximum shear stress (Figure 4.9), clearly a necessary condition if the material is to be economically used, then the working point must be point

$$C\left(-\frac{f}{2}, \frac{f}{2}, 0\right).$$

Since point C corresponds to pure shear with a shear stress $f/2$ the strain energy per unit volume is

$$\tfrac{1}{2}\ \text{(shear stress)} \times \text{(shear strain) or}$$

$$\frac{1}{2}\left(\frac{f}{2}\right) \times \frac{f/2}{G} = \frac{f^2}{8G} = \frac{(1+\nu)f^2}{4E} \qquad (4.9)$$

where E = Young's modulus, G = shear modulus and ν = Poisson's ratio. Expression 4.9 may be taken as a figure of merit for a material for use in a spring so long as the volume used is of overriding importance. If cost of material is most important, the figure of merit becomes

$$\frac{(1+\nu)f^2}{4Ec}$$

where c is the cost per unit volume.

We might consider whether point C is the best working point, and this will lead to another useful measure. It is readily shown that at points A and B of Figure 4.9 the strain energy per unit volume u is

$$\frac{(1-\nu)f^2}{E} \text{ and } \frac{f^2}{2E}$$

respectively, so that if $\nu = 0.3$, a common value for metals, the relative values of the u's are

A	B	C
2.15	1.54	1.00

It is scarcely possible to work the material of a one phase spring at point A, but there is no great difficulty in working it at B, e.g. in a multiple leaf spring.

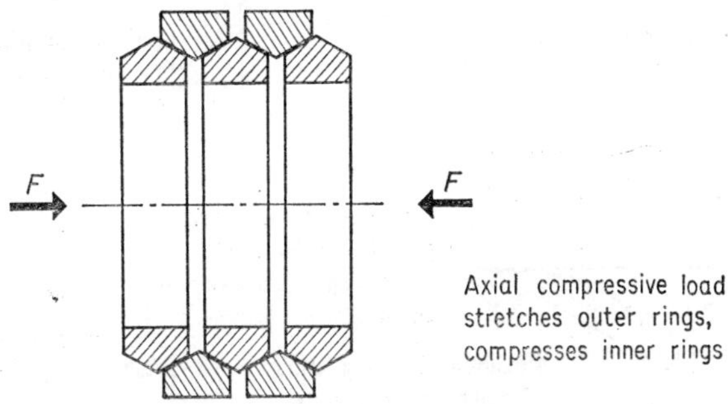

Figure 4.10 Ring spring

However, what matters is not the maximum strain energy per unit volume at the most highly stressed point, but the average strain energy throughout the material. Defining the *efficiency of use* of the material η_u as the ratio of the average to the maximum strain energy per unit volume, for the cylinder in torsion η_u is $\frac{1}{2}$, whereas for the rectangular cross-section beam in bending it is only $\frac{1}{3}$. This brings the multiple leaf spring and the helical compression spring about level as regards average strain energy per unit volume.

Much higher values of η_u are possible in some forms of spring. A hollow torsion bar will give η_u's of 0.80 or so, and a particularly high average strain energy may be had in the ring spring (Figure 4.10) resulting from a high η_u combined with working point B and its compressive equivalent. Because of the sliding friction between the rings, the ring spring has a hysteretic behaviour which limits its application but can be advantageous.

On the other hand, a simple uniform-section single leaf spring has an η_u of only 1/9; not only is the average strain energy only one-third of the maximum over any section, but the maximum itself varies from section to section as the square of the bending moment. This can be overcome by narrowing the breadth of the spring in proportion to the bending moment as in Figure

4.11(b) so that the maximum stress in all sections is the same. The result is a rhombic form, which in vehicle leaf springs is effectively cut into strips which are stacked one above another as in Figure 4.11(c). Notice that vehicle springs do not work quite like the rhombic spring—apart from anything else friction forces between them prevent them working entirely as independent beams—but the efficiency of use of material is nearer the ideal 1/3 than the uneconomical 1/9 of the spring of Figure 4.11(a).

Suppose it is required to suspend a tank of liquid nitrogen within an outer container; what is an appropriate figure of merit for the material of the suspension if it is required to keep the heat leakage into the tank a minimum?

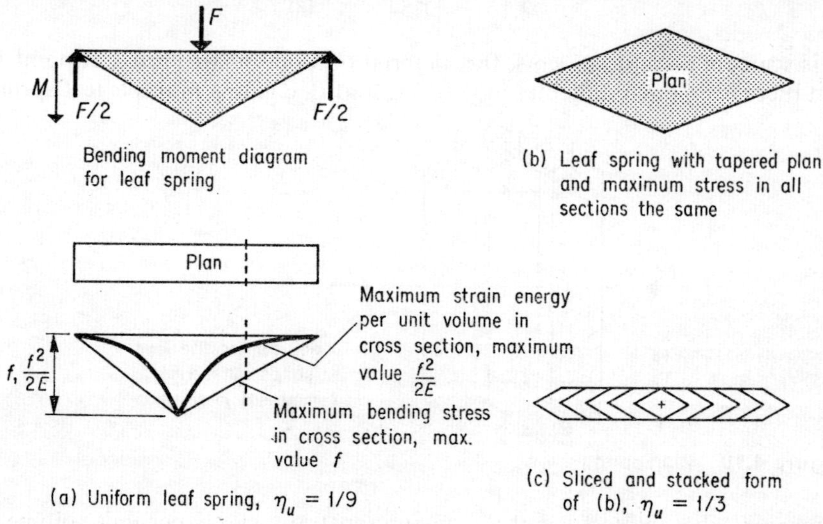

Figure 4.11 Material use in leaf spring

The immediate reaction might be to think of conventional insulating materials, but the figure of merit is k/f, where k is the thermal conductivity and f the allowable stress; if the material is stronger we can use a thinner section which will conduct less heat. Such materials as stainless steel and Inconel will rate well on this basis.

One last and rather long example of useful measures will be given. A ritual piece of information which is instilled into engineering students is the definition of the coefficient of performance of a refrigerator as the heat extracted from the thing cooled divided by the work expended. However, heat extracted is a very poor measure of the task performed since it does not take account of the temperature difference through which that heat must be 'pumped'. A performance ratio of 10 in a desalination plant would be poor, and one of 0.01 in a helium liquefier remarkably good.

A better measure of the performance is the thermodynamic efficiency η, defined as the ratio of the work required by a reversible machine performing the same task to the actual work. If we call the work required by the reversible machine the *ideal work*,

$$\eta = \frac{\text{ideal work}}{\text{actual work}} \text{ for a refrigerator or a heat pump,} \qquad (4.10)$$

and, by extension, for an engine

$$\eta = \frac{\text{actual work}}{\text{ideal work}} \qquad (4.11)$$

where the ideal work is that obtainable from a reversible machine using the same heat supply or the same fuel.

The ideal work required to effect a change in one unit of a substance (e.g. liquefy nitrogen) in a steady flow process, say from state 1 to state 2, is given by

$$b_2 - b_1 = \Delta b$$

where $b = h - T_0 s$, h is specific enthalpy, s is specific entropy and T_0 is the temperature at which heat is rejected, which is generally the ambient temperature. The function is called the availability (strictly the steady flow availability function), and the increase in availability Δb is a *measure* of the thermodynamic *task* to be performed.

Suppose now we imagine the change effected in two stages, from state 1 to state M and then from state M to state 2, the actual works required being W_{1M} and W_{M2}. Then there is an efficiency of each stage, and we can put, say

$$\eta_{1M} = \frac{b_M - b_1}{W_{1M}} \text{ and } \eta_{M2} = \frac{b_2 - b_M}{W_{M2}}$$

Clearly the higher the η's of the individual steps, the higher the overall η and the lower the work consumption of the process.

With these two useful concepts, thermodynamic efficiency and availability, at our disposal, let us look at the problem of liquefying nitrogen initially at room temperature and pressure, a task involving an increase of availability of, say, 100 units. Then it seems a reasonable plan to look for a practical process, *any* practical process, by which we can increase the availability of the nitrogen very substantially and as efficiently as possible. There are two such processes

(a) raising the nitrogen to a high temperature,
(b) compressing it highly.

Now since we seek to liquefy the gas (a) seems rather unpromising, which leaves (b). We can compress the nitrogen with a thermodynamic efficiency of about 0.80, and raise its b by about 60 units, which takes

$$\frac{60}{0.80} = 75 \text{ units of work}$$

to drive the compressor. The remaining increase of b of about 40 units can be achieved by cooling with a suitable refrigerator of one of the known sorts, which might have an efficiency of about 0.5. Thus the total units of work required are

$$75 + \frac{40}{0.5} = 155$$

giving an overall thermodynamic efficiency of

$$\frac{100}{155} = 0.65$$

If we liquefied the nitrogen entirely by cooling with a refrigerator of efficiency 0.5 we should need 200 units of work.

This deceptively simple argument reaches conclusions which are correct and might be valuable were there any need to liquefy nitrogen in large quantities, and it is a line of reasoning which is very easy to find once the two concepts on which it is based have become slightly familiar. Such concepts are valuable tools for the designer.

4.6 Bounds and limits

In the elementary theory of oil-lubricated bearings, there comes a point where it seems that the load capacity rises with the speed. As the engineer is likely to want to increase the speed of machines and the loading of their bearings at the same time, such behaviour seems very convenient and so very unlikely. We should at once look for the limit to this happy state of affairs, implicit in the parameter usually written ZN/P, where Z is the lubricant viscosity, N the speed, and P the specific loading of the bearing; if we increase N and P proportionately, the value of the parameter remains unchanged so long as there is no change in Z. But, as N and P are increased, the temperature in the vital area of the lubricating film will rise, causing a relatively very sharp fall in Z. Eventually a point is reached where the product ZN, and hence the load capacity, falls as N increases.

The engineer can perhaps be forgiven a little anthropomorphism, if he thinks of the natural order as imbued with cussedness, never giving with one hand except to take away with the other. He rarely makes an advance in one direction except at the expense of losses in another, but it is this very sort of difficulty that gives his work its challenge. An important aspect of insight into an engineering problem is a clear and balanced view of the conflicting constraints set to its solution by natural laws, available materials and the state of the art.

Consider the problem of designing a beam AB (Figure 4.12) of minimum weight, subject to a given bending moment M and shear force S, which are specified functions of x, the distance from A. The allowable stresses are $f/2$ in pure shear and f in simple tension or compression, corresponding to the maximum shear-stress criterion used elsewhere (Figure 4.9). If the depth of the beam is h and it consists of two flanges, of cross-sectional area A_F each, which take the bending moment, and a web of area A_W that takes the shear,

$$A_F = \frac{M}{fh} \text{ and } A_W \simeq \frac{2S}{f}$$

and the total volume of material

$$\int_A^B (2A_F + A_W)\,\mathrm{d}x = \frac{2}{fh}\int_A^B M\mathrm{d}x + \frac{2}{f}\int_A^B S\mathrm{d}x \qquad (4.12)$$

This analysis will do fairly well as long as the beam is shallow and the structural loading coefficient is high enough and the material stiff enough for no questions of stability to arise—put another way, as long as the beam is of

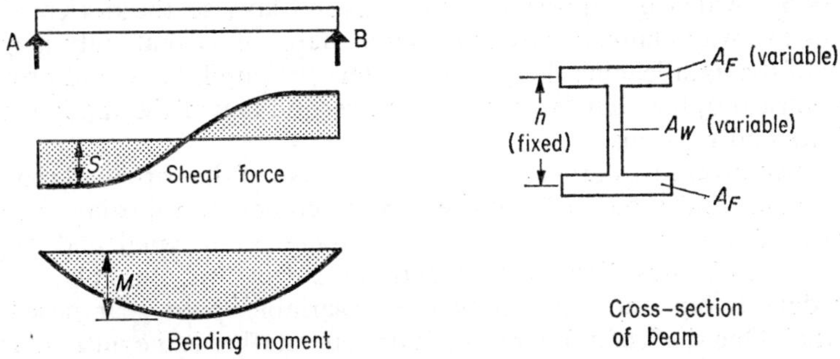

Figure 4.12 Structural task of a beam

high solidity. But if we attempt to minimize expression 4.12 with respect to h it becomes clear at once that its validity must be limited, since the first term in it decreases indefinitely as h increases, and the second is independent of h. Expression 4.12 has no minimum, but the engineer's view of the physical world tells us there must be one; hence expression 4.12 must be an inadequate representation of a deep beam.

There must be a solution to the problem of the minimum-weight beam, and a sufficient number of particular solutions of this sort of problem are known for us to estimate the volume of material in it fairly closely. This volume we shall call 'ideal', by extension of the use of 'ideal' in thermodynamics. Normally minimum-weight structures are highly impracticable (not in the case of the disc; see section 3.11) but practicable designs can be found which are not very much heavier. The ratio of the volume of material in a practical design to the ideal volume is a good measure of its excellence as regards economical use of material. Moreover, a study of the 'ideal' form will provide a great deal of help in designing a good practical form [17].

The most famous limitations in engineering science are the first and second laws of thermodynamics; the statements of the second law as an embargo on perpetual motions of the second kind and as Carnot's theorem are indeed expressly limitations of possible engineering devices. The scope of the second law is not always realized; consider the following proposal to enable soldiers to see better at night.

'Make field glasses with object glasses ten times the diameter of the pupil of the eye (about 0.2 in.) and a magnification of five times. The object glasses will collect one hundred times as much light as the naked eye, say, ninety times allowing for absorption and reflection by the glass, but only produce an image on the retina twenty-five times the area, and so $90/25 = 3.6$ times as bright.'

If it were possible to construct such an optical system, an image of the Sun could be formed which would be brighter than the Sun itself, and with suitable insulation of the area on which it was formed, would become hotter than the Sun. We should then have heat flowing of its own accord from a colder to a hotter body, which is impossible by the second law, so the device will not work (in the way claimed, anyway). What happens is that only a quarter roughly of the light leaving the eyepiece enters the pupil. We could prove this by geometrical optics, but the proof by the second law of thermodynamics is immediate and rigorous.

In the past inventors pursued the will of the wisp of the perpetual motion of the first kind, and modern favourites are machines that disobey Newton's third law of motion that to every action there is an equal and opposite reaction. The obvious form is the reactionless drive (or, possibly, the anti-gravity device) but another variety is the variable-speed gear based on a differential. One shaft of a differential is driven, another is the output, and the third is allowed to run away, as it were, but only in a controlled fashion. This is a failure of insight, an inability to recognize that the torque in one shaft determines those in the others and so a wide speed range can be achieved only with a large power flow in the third shaft. To feed this power into the output we should need a variable-speed gear, which is what we started out to design.

Nevertheless, this is not a fruitless train of thought. Suppose an automatic transmission for a car needs a power throughput of 50 kW and a ratio of top to bottom gear, what we shall call the range of ratio, R, of 4. If we use a hydrostatic transmission, the pump has to be capable of absorbing 50 kW, which is a fair measure of its size and cost. If at maximum power in bottom gear the hydraulic motor speed is ω and the torque it develops is T, then it has also to run at 4ω at maximum power in top. Since $\omega T = 50$ kW, the product of the maximum torque T and the maximum speed 4ω is 200 kW. This 'power', called the 'corner power,' is another useful concept. Although the true power never exceeds 50 kW it is nearly as exacting to produce a motor with a product of maximum torque and speed of 200 kW as one with a true power of 200 kW. The corner power is a good rough measure of the hydraulic task and hence the amount of machinery and its cost. Thus the whole transmission requires about 250 kW-worth of hydraulics.

Let us take the third shaft of our original differential and connect it via a hydraulic variable-speed gear (v.s.g.) to the output shaft (Figure 4.13). We now have an arrangement in which the power is divided in the differential, flowing through two branches A and B to be recombined in the output shaft. As only a part of the power flows through the branch with the hydraulic elements, the corner power is reduced. For the same reason, the variation in the overall ratio of the whole system is less than that in the hydraulic 'branch', but the range of ratio of the hydraulic variable-speed gear is infinite and so a reduction can be tolerated. In fact, the whole arrangement may be regarded as 'trading off' the excess range of ratio available in the hydrostatic elements for increased throughput. The question is, just how low can we make the required corner power of the hydrostatic elements in this way? What is the 'conversion rate', as it were, for exchanging excess range of ratio for higher throughput?

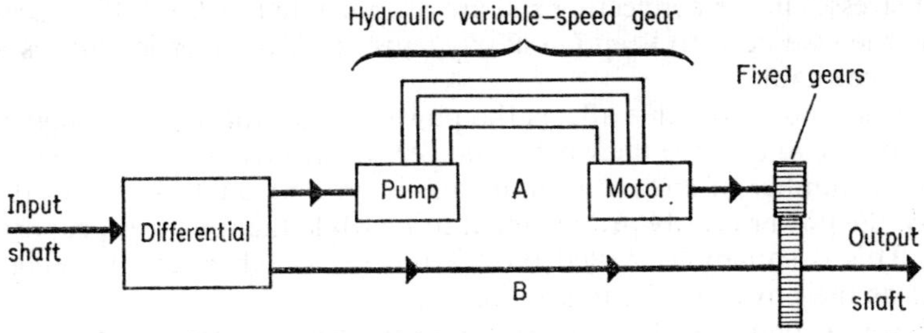

Figure 4.13 Variable-speed gear with differential

The answer in the case of the car transmission is disappointing: without reversal of power flow in the A branch (which is awkward and inefficient), we cannot reduce the corner power of the hydraulic elements below 180 kW, which is not enough to make the scheme viable. On the other hand, in an application which only needed a range of ratios R of 1.5, such a scheme could reduce the corner power requirement by two thirds, a very worthwhile saving. The split drive can reduce the total corner power in the ratio $(R - 1)/R$ at most; this restriction applies to 'input and output summer' arrangements, to the type reported in reference 18, and to much more complicated schemes also.

Such 'split-drive' transmissions have found application in feed pumps, constant-speed drives for aircraft generators and in a few other places. They are limited in a quite rigorous way, so that we can say at once that they can do little good in a car transmission, but they may effect large savings where only a small range of ratios is required.

This is an example of a highly specialized limiting law governing a particular design problem. It not only tells us that the split drive transmission requires a certain minimum corner power, but if we study the conditions under which it applies, it suggests how rather similar mechanisms free from its own limitations may be constructed.

4.7 Scale effects

In many problems a useful way of acquiring insight is to study the effects of scaling. If all the dimensions of a pressure vessel are increased by the same scale factor (including wall thicknesses) then the stresses are unaltered. If all the dimensions of a gravity-loaded structure such as a suspension bridge are scaled up, the stresses increase in the same proportion.

Consider an internal-combustion engine, which runs with angular velocity ω and has a crank throw L. Then the centripetal acceleration of the crank pin is $L\omega^2$, and all the accelerations of the moving parts are proportional to $L\omega^2$. If we take L as a typical dimension, then the mass of the connecting rod, for example, is proportional to L^3 and so the inertia load it exerts on the bearing is proportional to $L^4\omega^2$. The bearing area is proportional to L^2, so that the specific loading of the bearing is proportional to $L^2\omega^2$. In a similar way, the inertia stresses in the connecting rod are proportional to $L^2\omega^2$. If we scale the engine, then we need to keep $L\omega$ nearly constant if the inertia stresses are to remain acceptable.

Let us see now how this affects the power/weight ratio of the engine. The swept volume and hence the work done per stroke increases as L^3, but ω and hence the number of strokes in unit time is proportional to $1/L$, so that on balance the power is only proportional to L^2 while the weight is proportional to L^3. This is sometimes called the '$1\frac{1}{2}$ power law', because the weight increases as the power to the index 1.5.

One way round the $1\frac{1}{2}$ power law is to increase the swept volume by increasing the number of cylinders without making them bigger. Further, since the power for a fixed maximum speed (fixed ωL) depends only on piston area, the weight per unit power is inversely proportional to the stroke. Logically, then, to make a light engine, we make the stroke as short as may be relative to the bore ('over-square') and use a large number of small cylinders, a philosophy embodied in the original B.R.M. engine.

The case of a turbine is similar. The mass flow rate of fluid, and hence the power, is proportional to the square of the diameter. If all the dimensions are scaled up equally, the power is proportional to L^2 and the weight to L^3 and we have the same $1\frac{1}{2}$ power effect. But we have rather more freedom than this. Why need we scale up the axial dimensions by the same factor as the diametral ones, indeed, why need we scale them up at all?

The answer lies in the bending stresses in the blading. Imagine that we scale all the transverse dimensions (including the diameter and all the blade lengths) by one factor and all the axial dimensions (including the blade chords, disc thicknesses, overall length etc.) by another, and let D and L be typical transverse and axial dimensions. The weight is proportional to D^2L. The projected area of a blade surface is proportional to DL, and since the pressure difference over it is fixed, the load on it is proportional to DL also. The bending moment at the fixed end of the blade is proportional to D^2L, and the section modulus there is proportional to L^3. The bending stress in the blade is proportional to D^2/L^2 and so remains unaltered if we scale the axial dimensions equally with the transverse ones. Thus the $1\frac{1}{2}$ power law will be followed as long as the axial dimensions are determined by stressing, and the ratio of weight to power will be proportional to the square root of the power, as in the curve EBC of Figure 4.14.

But suppose we scale down a turbine, say, represented by point C in Figure 4.14. The axial dimensions (the blade chords included) are scaled down proportionately with the transverse ones until a point is reached where it becomes impracticable to reduce the smallest ones any more. From this point (B in Figure 4.14) onwards, only some of the axial lengths can be scaled down, and so the curve drifts away from the parabola EBC. By the time even the largest blade has its chord determined, not by stress, but by the smallest size it is simple to make, no further axial scaling is possible and the weight/power ratio on this simple analysis will remain constant.

Of course, these considerations only apply to the blading and other parts governed by the blading, like discs. The casings where they are determined in

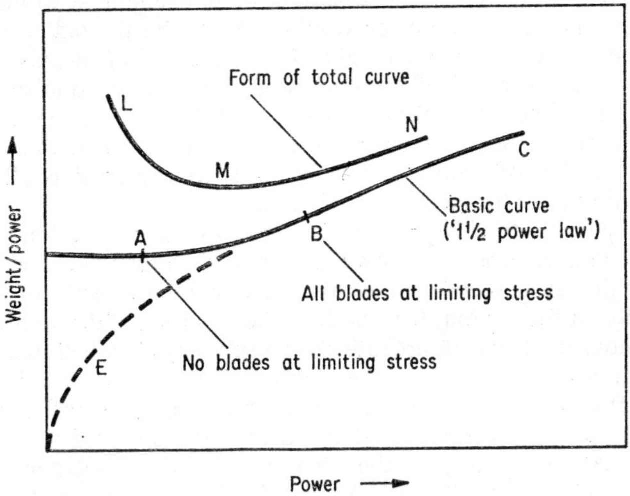

Figure 4.14 Specific weight of aircraft gas turbines

thickness by pressure stresses will thin proportionately with the diameter and so will march with the blades. In small sizes the thickness may be nearly independent of diameter, and so the casing weight/power ratio will eventually increase with decreasing power. At sizes above point B in Figure 4.14, the parameter measuring the difficulty of providing sideways stiffness in rotors, the x axis in Figures 2.4–8, will remain constant, but below B it will increase because the length decreases less rapidly than the diameter. Proportionately more material will be needed for transverse stiffness, and this effect also will curl the curve upwards. Many accessories (and many details such as bearing assemblies) will not decrease in weight proportionately with the power. When all these are taken into account, the form of curve obtained is something like LMN in Figure 4.14, with a minimum at M.

If the centrifugal stresses are limiting, an electrical machine behaves differently under scaling. If the flux and current densities are kept the same, the current taken and the total flux each increase as L^2. The centrifugal stresses are unaltered if the surface speed remains the same, so the angular velocity is proportional to L^{-1} and the e.m.f. is proportional to the total flux times the angular velocity, or L. The power is proportional to the e.m.f. times the current, or L^3. This simple result is slightly complicated by the field windings not following quite the same law, a matter we shall not discuss, and also by the problem of cooling, which is worth studying here.

In the scaled machine, the copper losses per unit volume are unaltered; if the lamination thickness is scaled, the iron losses per unit volume are also unchanged. The total heat evolved within the materials is proportional to L^3, but the surface available for cooling is only proportional to L^2. Very roughly, if no compensating measures were taken, the temperature rise would be in proportion to L. If it were acceptable on a small machine, it would not be so on a bigger one, and the power would have to be limited to an L^2 law. Maintaining the power proportional to L^3 depends on the provision of better cooling. At the lower end of the size range this can be done simply by increasing the local heat transfer coefficient over the areas naturally available, by forcing air past them at higher speeds, but in large machines extra area is needed, ideally, of course, sufficient to keep the total cooling surface proportional to L^3. This means a fixed passage size.

A rough parallel to this problem is found in the respiration of animals. Very small creatures are able to respire directly through the cell walls. Insects have branching systems of ventilating tubes opening directly to the outside which supply air throughout their bodies; the pumping of the air is effected by muscular movements which operate on the tubes as if on bellows. The much larger mammals have an even more elaborate system, using an intermediate fluid (blood) with its own pump (heart) as well as a large surface area in the lungs and an air pump (rib cage, diaphragm and associated muscles). It has been suggested, very plausibly, that there are no large insects in spite of the great success of this versatile basic design of animal because of the limitations imposed by their respiratory system. The problem here, however, is more than that of keeping the surface area proportional to L^3: it involves the work required to pump the air in and out of the system, and the space occupied by the tubes (see Question 4.9). As the insect is made larger, the amount of air flowing through unit volume of its body to reach more interior points increases in proportion to L. Similar problems arise in town traffic [19] and nuclear-reactor cores.

As a simple example of this last sort of problem, consider heat being evolved in a long rod of diameter D at a constant rate per unit volume and extracted at the surface (Figure 4.15). The total heat flow per unit length of rod is a proportional to D^2, and the area of its flow path in the material to D, so that the heat flux density, and hence the temperature gradient, is proportional to D. Thus the temperature difference between the centre and the surface of the rod is proportional to D^2.

In Chapter 6, the design of gears will be considered, and it is convenient to study the dimensional aspects here. The loads we can apply to gear teeth are limited by bending stresses and surface stresses. If we scale a gear tooth up, the modulus of its root section increases as L^3, the arm of the tooth load about it increases as L, and so the load which can be carried increases as L^2. The torque on the gear can thus be increased as L^3.

The radius of curvature of the tooth profile is proportional to L, and so therefore is the load per unit length of facewidth that can be carried. The facewidth is proportional to L, so once again we have a load capacity proportional to L^2 and a torque capacity proportional to L^3. In this case we can scale the length L and the diameter D independently and find that the torque capacity is proportional to D^2L, which is just $4/\pi$ times the volume of the pitch cylinder. Indeed, we can write for a pinion of high quality.

$$\text{allowable torque} = \frac{CD^2L}{1+k} \tag{4.13}$$

where C is a constant and k is the number of teeth in the pinion divided by the number of teeth in the gear. The writer has used $C = 300$ p.s.i. (2 N mm^{-2})

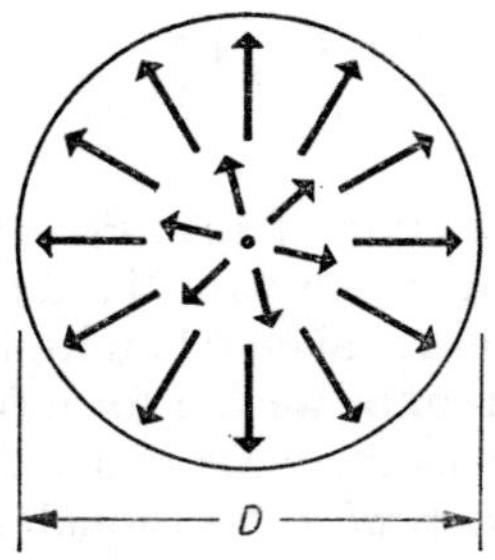

Figure 4.15 Flow of heat outwards in a rod

roughly for case hardened and ground En 39B in aero-engines; at very high surface speeds, where design has to be modified from what would otherwise be the optimum to avoid scuffing, C is reduced somewhat, as in rocket pump drives [20].

4.8 Dimensional analysis and scaling

The reader who has a knowledge of dimensional analysis will perhaps wonder why it has not been used here. The reason is that scaling ideas are a more powerful and sophisticated device, and, being capable of direct physical interpretation, are both more easily understood and more useful for gaining insight.

Everything that can be done by formal dimensional analysis can be done by scaling, but a very great deal which can be done by scaling cannot be done by dimensional analysis. On the other hand, dimensional analysis is quicker, surer, and tidier so far as it goes. We should use whichever fits the situation better (and that sometimes means both).

The advantages of scaling are that

(1) we can tackle problems in which there is not strict similarity, and
(2) we can use physical reasoning.

In the usual treatment of pipe flow by dimensional analysis, physical reasoning is used when it is assumed that for fully developed flow the pressure gradient along the pipe is constant. The difference between the approaches with and without physical reasoning is brought out by the following example.

A roller of radius R, length L is pressed against the edge of a semi-infinite plate, thickness L of the same material by a uniform force per unit length P/L. What is the maximum stress σ_M and the halfwidth b of the contact strip?

This is a simple form of Hertz's problem. By dimensional analysis we can only say: the parameters involved are P, R, L and E, the Young's modulus of the material, and ν, Poisson's ratio. The effect of ν is nearly always small and we ignore it. Then

$$\sigma_M = \frac{P}{RL} \mathrm{f}\left(\frac{R}{L}, \frac{P}{ER^2}\right) \tag{4.14}$$

where f is some unknown function of R/L, P/ER^2. This equation is obtained by putting σ_M equal to some group with the dimensions of stress (we have chosen P/RL, but we could have taken E) multiplied by a function f of all the independent dimensionless groups we can form from the parameters P, R, L and E. The 'π-theorem' tells us there are two such groups.

Now strictly speaking this is *all* dimensional analysis can tell us. But by physical reasoning, we expect the behaviour in all sections perpendicular to

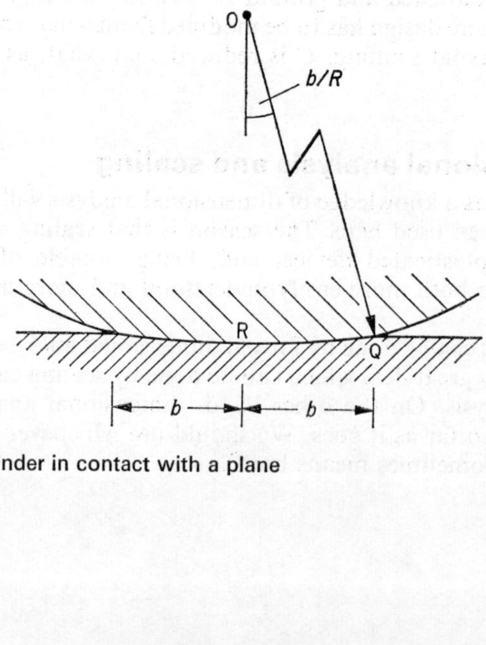

Figure 4.16 Cylinder in contact with a plane

the axis of the cylinder to be the same, so that P and L will only appear as P/L, the load per unit length, and instead of equation 4.14 we can write

$$\sigma_M = \frac{P}{RL} g \left(\frac{P}{ERL}\right) \tag{4.15}$$

where g is an unknown function of the single parameter, P/ERL.

To proceed further, we use Figure 4.16 and scaling. If Q is a point at the edge of the contact flat, the two surfaces there are parallel to each other after loading, while before loading they had an angle b/R between them. The sum of the rotations of elements at Q must therefore be b/R. Now the rotation of an element of a strained body relative to some other element (e.g. one at R) is proportional to the strains in that body so that σ_M/E, which is a strain, is proportional to b/R.

The total load P is proportional to the area of the contact flat times a typical stress, or

$$P \propto \sigma_M \,.\, bL \propto \frac{bE}{R} \,.\, bL, \; b \propto \sqrt{\left(\frac{PR}{EL}\right)}$$

$$\sigma \propto \frac{bE}{R} \propto \sqrt{\frac{PE}{RL}} \tag{4.16}$$

Now this is as far as we can get, but it is a good deal further than equation 4.15. We know that

$$g\left(\frac{P}{ERL}\right) = K\left(\frac{P}{ERL}\right)^{-\frac{1}{2}}$$

and only the constant K is undetermined. In fact it is about 0.42 for $\nu = 0.3$.

This example has little to do with design, but it shows very well the power of the extended form of dimensional analysis, using physical reasoning, which is here called 'scaling'.

4.9 Proportion

There is much truth in the adage, 'if it looks right, it is right', and perhaps even more in the converse, 'if it looks wrong, it is wrong'. A good designer has the sort of feel for proportion which enables him to look at a drawing and spot the weaknesses or the inappropriate size of certain features. A simple example will show how this can be so.

Suppose a pinion meshes with a single wheel and is mounted in roller bearings at each end. If, as will usually be the case, the principal function of the bearings is to carry the reaction to the tooth load on the pinion, this consideration will fix the relative sizes of the components, so that even a slightly practised eye can recognize wrong proportions. The load-carrying capacity of the pinion tooth is proportional to the facewidth and the relative radius of curvature at its point of contact with the wheel (section 4.8) and the radius in question is in turn roughly proportional to the pitch diameter (say, about 1/6 of it in a well-designed pinion meshing with a wheel with about three times the number of teeth).

Similarly, the load-carrying capacity of a roller bearing with a given number of rollers (say, 11) will be proportional to the facewidth (the length) of the rollers and the relative radius of their curvature at their point of contact with the outer race, which is about 1/12 of the pitch diameter of the roller cage. A common assumption is that the heaviest loading on a single roller is about $z^{-\frac{2}{3}}$ of the total load, where z is the number of rollers. If $z = 11$, we can regard the load as shared among $11^{\frac{2}{3}}$ or, say, 5 rollers. If the pitch diameters of the roller cage and the pinion are about the same, and if the pinion is case-hardened and ground, so its surface strength is comparable with that of the rollers, the ten effective rollers in the two bearings, having half the relative radius of curvature of the gear teeth, are suitable for carrying the tooth load on a pinion facewidth $10/2 = 5$ times the roller width. If there are more than 11 rollers, the ratio will be lower in the proportion of $z^{-\frac{1}{3}}$ roughly.

4.10 Change of viewpoint

A most useful trick for increasing insight, but one rather difficult to illustrate briefly, is changing the viewpoint.

Suppose we are designing some bathroom scales; the suspension of the platform must be such that the indicated weight is the same wherever the

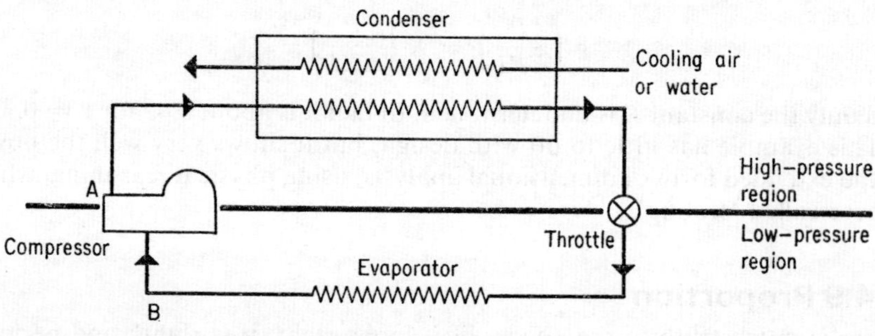

Figure 4.17 Vapour compression refrigerator

person stands on it. When a load W is placed on the platform there is a load kW in a certain link of the mechanism; the requirement is that k should be a constant and not vary with the position of W on the platform. This is a statical condition and it can be designed for.

If we imagine the link given a small movement h along its length, then the point on the platform where W is situated moves s, say, in the vertical direction. Since the mechanism is almost frictionless, we can apply the principle of work and put

$$hkW = sW, \; s = hk$$

Now since k is to be constant and independent of the position of W, s must also be independent of the position of W, i.e. the platform must remain parallel to its original position. This is a kinematic condition and it can also be designed for, and because mechanisms which satisfy it can be visualized relatively easily, this changed view of the problem is more likely to be fruitful. Most readers will be able to think of various parallel motions which might be adaptable to use as a platform suspension, and the easier visualization of the kinematic property may lead to less obvious forms suggesting themselves.

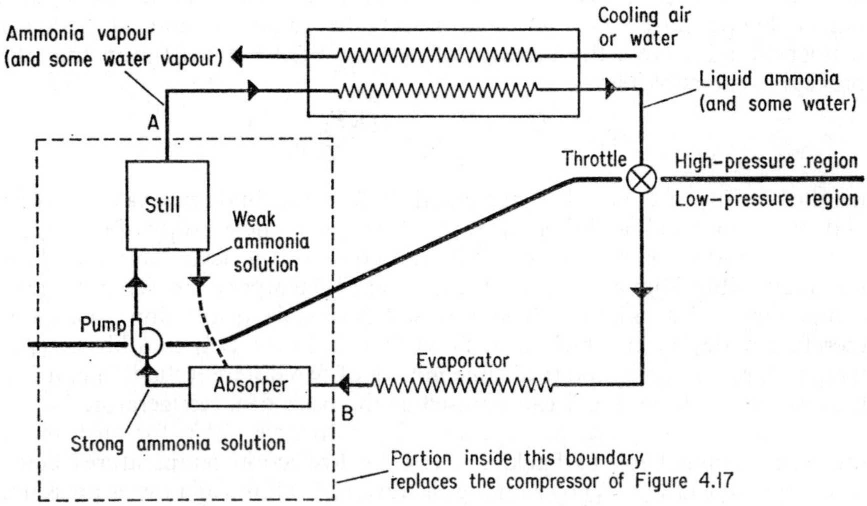

Figure 4.18 Absorption refrigerator

A good example of change of viewpoint is given by the absorption refrigerator. As a reminder, Figure 4.17 shows the flow diagram of an ordinary vapour compression refrigerator. To fix ideas, the refrigerant is taken to be ammonia, which is compressed and condensed at roughly ambient temperature and the higher pressure in the condenser. The resulting liquid ammonia is passed through the throttle, yielding mostly liquid and a little vapour at a lower temperature and pressure. The liquid is then evaporated in the evaporator, extracting heat from whatever it is required to cool.

Figure 4.18 shows the flow diagram of an absorption refrigerator working on ammonia and water. From A to B it is identical with the vapour-compression cycle. Instead of using a compressor, however, the ammonia vapour leaving the evaporator at B is absorbed in a weak solution of ammonia to give a strong solution of ammonia with a volume perhaps one per cent of that of the vapour. This strong solution is raised to the higher pressure by a liquid pump. Then the ammonia vapour is boiled out of the solution by the application of heat in a still, leaving a weak solution which is returned to the absorber. The available energy input comes mainly from the heat source used for the still, the liquid pumping term being nearly negligible.

Thus the absorption refrigerator can be viewed as a vapour-compression refrigerator in which the compressor is replaced by an absorber, liquid pump and still. This is already a fairly abstract view, and a helpful one. But we can do a great deal better.

A distillation column is, ideally, a reversible device [21]. In practice, of course, like all real devices, it is irreversible to a lesser or greater extent. Leaving aside certain irrelevant changes of enthalpy, what happens in a distillation column is that the fluids flowing through it take in heat Q at a higher temperature, T_1, say, and reject the same amount at a lower temperature T_2. Now the increase of entropy of the fluids (given that the process is reversible) is

$$\Delta s = \frac{Q}{T_1} - \frac{Q}{T_2} = -\frac{Q(T_1 - T_2)}{T_1 T_2}$$

i.e. the entropy of the fluids is decreased. In fact, the fluids enter as a mixture and leave separated, and their entropy decreases by the entropy of mixing.

Thus the ideal or reversible distillation column is a device in which heat flows reversibly from a higher (T_1) to a lower (T_2) temperature, separating the components of a mixture. The *reversed* reversible distillation column is therefore a device in which heat flows from a lower (T_2') to a higher (T_1') temperature, driven by the reversible mixing of two different fluids; i.e. it is a heat-pumping element and can be used as the basis of a refrigerator.

To make the complete absorption refrigerator, we put a forward and a reversed column back to back, the two heat-rejection temperatures being made the same ($T_2 = T_1'$) by running the reversed column at a lower pressure. The fluids A and B separated in the forward column drive by their mixing the heat-pumping reversed column, and a liquid pump is needed to return them afterwards to the high pressure forward column.

This last paragraph gives a highly abstract and fundamental view of an absorption refrigerator, much more powerful as a conceptual tool than the vapour-compression-with-compressor-replaced view. It makes available for improving the performance many general ideas in thermodynamic design and widens our view of what fluids A and B might be like. It increases our insight to a point where difficult questions that would be hard to answer with pencil and paper and plenty of time, can be resolved in the head.

It is hoped that readers whose knowledge of thermodynamics is not sufficient for them to understand this illustration fully will nevertheless be able to grasp its general nature. The majority of cases of changed viewpoint contributing to insight suffer from defects as examples; they generally require a great deal of background stuff and scene-setting, and the change itself is usually unimpressive and its value obscure to those not involved in the problem. The two chosen illustrations are unrepresentative in that they are short and tidy, and that the new view supersedes the old: more often the new view merely complements the old.

Two features frequently shown by changes of viewpoint are

(a) a higher level of abstraction, and
(b) a change of measures, especially to a *task* measure (section 8.1).

As an example of (b), it is frequently advantageous to regard toothed gearing as a torque-generating device rather than a speed-changing one. Indeed, throughout the design of gearing it is useful to alternate between the kinematic and the statical viewpoint, but this is getting close to the diversification of approach discussed elsewhere (section 1.5) rather than the development of insight.

Questions

Q.4.1(3). Figure 4.19 shows the principle of a suggested magneto-hydrodynamic submarine. A portion Z of the hull bears a circumferential pattern of alternating magnetic poles and electrodes. The magnetic poles produce the sort of field shown by the solid lines in the figure; current flows between the electrodes as shown by the dotted lines, the conductor being the surrounding sea-water, which is forced backwards, so propelling the submarine forward.

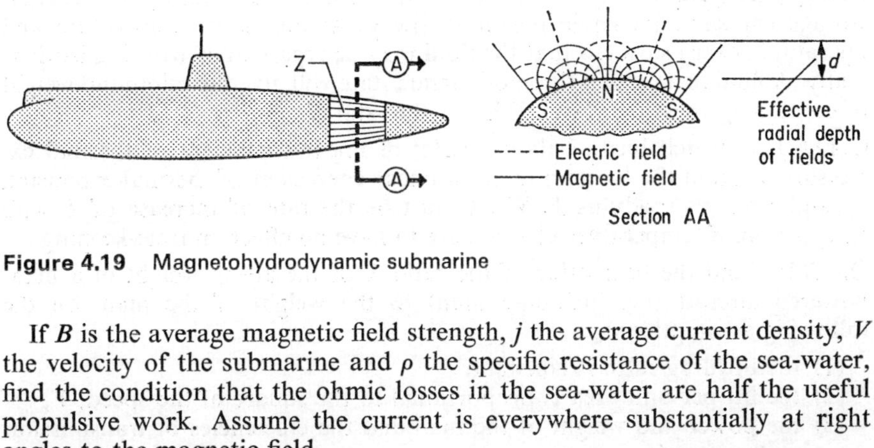

Figure 4.19 Magnetohydrodynamic submarine

If B is the average magnetic field strength, j the average current density, V the velocity of the submarine and ρ the specific resistance of the sea-water, find the condition that the ohmic losses in the sea-water are half the useful propulsive work. Assume the current is everywhere substantially at right angles to the magnetic field.

If $B = 1$ Wb/m^2, $V = 10$ m/s, $\rho = 0.2$ ohm m find roughly the radial distance, d, that the active zone must extend beyond the hull if this condition is to be met; the area of Z is 0.2 times the total surface area, and the drag may be taken as 1.5 V^2 N/m^2 of hull surface.

Q.4.2(1). How must Figure 4.7, illustrating the compressor shaft diameter problem, be modified to cope with cases where the shaft material is of a different density (ρ_1) from the disc material (ρ_2)?

Q.4.3(1). Propose, in algebraic form, figures of merit for materials for the following purposes:

(a) a slender solid strut, to be of least diameter,
(b) a flywheel, to be used as an energy store, of minimum volume,
(c) a fluid for use in a fluid spring of minimum volume,
(d) a fluid for use in a fluid spring of minimum weight,
(e) a material subject to high thermal stress due to fixed temperature differences.

Express your results in a form where high numbers are good, in terms of the following

E: Young's modulus K: bulk modulus
f: typical allowable stress ρ: density
α: coefficient of thermal expansion

Q.4.4(2). On the same lines as Question 4.3, propose figures of merit in these more difficult cases:

(f) a tape in a gear grinder, which is wound on and off a drum to form a mechanism equivalent to a rack and pinion. The requirement is the minimum stretch in the tape under a given load, the drum size being fixed.
(g) a spin-drier shaft, which must be able to withstand a certain given bending moment, under which its deflection should be as large as possible.

Q.4.5(3). A fluid spring is required of minimum weight and given energy-storage capacity. Its environment is always at the same temperature and operation is very slow, so that the fluid may be regarded as working isothermally. A fluid in a certain thermodynamic state will give the minimum weight of containing pressure vessel. What is that state?

Q.4.6(2). A watch has a balance wheel of linear coefficient of thermal expansion α_W, and the hairspring has a linear coefficient of thermal expansion α_H and Young's modulus E. What must be the rate of increase of E with temperature if temperature changes are to have no effect on time-keeping?

Q.4.7(3). Find the best value of the ratio K of the all-up weight of a man-powered aircraft (i.e. including man) to the weight of the man, on the following assumptions:

(1) the lift/drag ratio is constant,
(2) the lift per unit area is proportional to the square of the speed, V,
(3) the power and weight of the man and the efficiencies of transmission and propulsion are constant,
(4) the optimum shape of aircraft is independent of K.

Make in turn each of the following assumptions about structural weight:

(a) it is proportional to wing area, and hence to L^2, where L is a typical dimension,
(b) it is proportional to wing area and also to the square root of the load per unit area.

Q.4.8(4). On the same topic as the last question, what would be the optimum value of K on a planet with very dense air, a very high force of gravity, and very powerful inhabitants?

Q.4.9(1). Assuming that traffic in larger towns does not tend to become relatively more localized than in smaller ones, and that densities of population, social habits etc. are also not affected by the size of towns, how will the internal traffic density vary with N, the total urban population?

Q.4.10(2). A coin of radius r is placed on a horizontal surface. A similar coin is placed on top of it, slightly eccentrically so that it overhangs on one side. A third coin is placed on the second, and so on, till the pile contains $n + 1$ coins. What is the maximum amount by which the top coin can overhang the bottom one?

Q.4.11. Sketch a plane mechanism to apply a pure torque to a body. Find a kinematic property which any such mechanism must have and see if this suggests another design.

Q.4.12(2). The inlet to the last stage of a radial outward-flow steam turbine is the circumferential gap between a disc and a ring, both 1200 mm in diameter and spaced apart axially at a distance of 600 mm. The blades are to be supported by the disc at one end and the ring at the other, and the whole rotates at 3000 rev/min.

Each blade forms a bridge between the ring and the disc, withstanding, not gravity, but centrifugal forces. Find the equivalent length of gravity-resisting bridge and so the form the blade should take.

Answers

A.4.1. $2j\rho = BV$; $d = 30$ m. A unit cube of water in the active zone has a current j in it, is in a field of strength B, and so has a force jB on it. The *useful* work of propulsion per unit time is thus jBV (see section 5.2). The ohmic loss per unit volume is ρj^2. If H is the hull area, 0.2 Hd is roughly the active volume, and 0.2 $HdBj$, the propulsive force $= 1.5\ HV^2$.

$$\text{Hence } dj = 750 \text{ A/m}$$

Also, since $2j\rho = BV$, $j = 25$ A/m², $d = 30$ m. This is clearly impracticable, so it does not matter that 0.2 Hd = active volume has broken down.

A.4.2. Substitute $\dfrac{\rho_1 p}{\rho_2 t}$ on x axis.

A.4.3. (a) E. (b) f: the kinetic energy stored in unit volume of disc is $\frac{1}{2}\rho V^2$ but the allowable ρV^2 is proportional to f (see section 3.11). (c) $1/K$. (d) Depends on containment also, but see Q.4.5. (e) $f/E\alpha$.

A.4.4. (f) f: if a material of half the value of E were used, the stresses produced by bending round the drum would be the same if the tape were twice as thick, and hence of the same stiffness in tension. (g) $f^{4/3}/E$: the bending strength requirement means that d, the shaft diameter, is proportional to $f^{-1/3}$. The strain in the shaft under a given deflection is proportional to d, and so the stress is proportional to dE.

A.4.5. Saturated: the figure of merit is $1/K$, which is infinite for a saturated vapour, isothermally compressed. The pressure is constant, and the change in volume can be practically the entire volume of the pressure vessel. The corner work is equal to the real work (sections 5.8–5.9).

A.4.6. $(2\alpha_W - 3\alpha_H)E$: the fractional increase in the radius of gyration of the balance wheel for unit temperature rise is α_W, so the increase in radius of gyration squared and hence the moment of inertia is $2\alpha_W$ (Binomial theorem). This must be balanced by an equal fractional increase in the torsional stiffness of the hairspring, which increases by $4\alpha_H$ because of the increase in section second moment of area, but decreases by α_H because of the increase in length.

A.4.7. (a) $1\frac{1}{2}$. (b) 2: since the lift/drag ratio is constant, the drag is proportional to K and the propulsive power to KV. In each case KV must be minimized.

(a) Lift $\propto K$, lift $\propto L^2V^2$, structural weight $\propto L^2$, structural weight $\propto K - 1$. Hence $K \propto L^2V^2$, $L^2 \propto K - 1$,

$$V^2 \propto \frac{K}{K-1},\ KV \propto \frac{K^{\frac{3}{2}}}{\sqrt{(K-1)}}$$

Minimize KV and $K = 1\frac{1}{2}$. Similarly for (b).

A.4.8. 1.50–1.67: the wing loading will be high, the structure loading coefficients high, and so the structure will be fully stressed by the loads it carries. In such a case the section areas will be proportional to K, and the structure weight, therefore, to KL.

A.4.9. Proportional to $N^{\frac{1}{2}}$: the area is proportional to N, and hence a typical length of journey to $N^{\frac{1}{2}}$. The product (traffic units × average distance travelled) is proportional to $N^{3/2}$, and so the amount passing through any unit area is proportional to $N^{\frac{1}{2}}$. This is like the insect breathing problem, but easier.

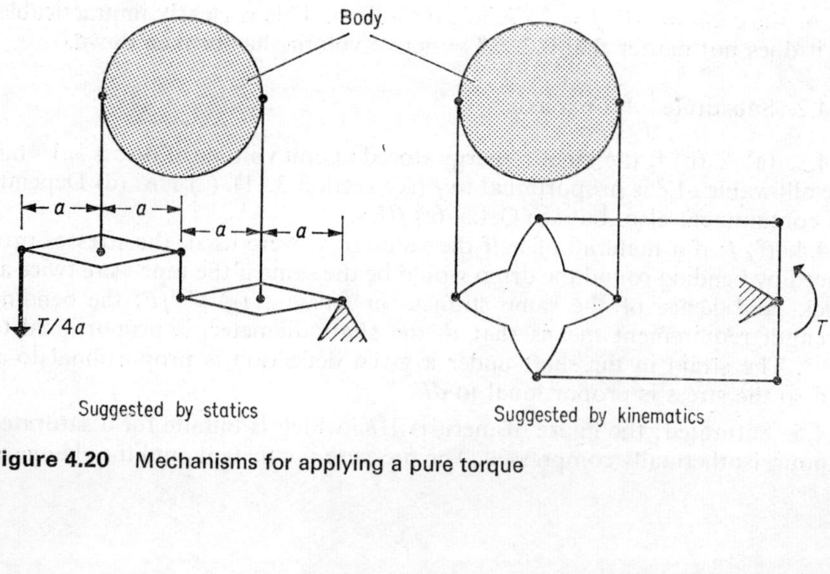

Figure 4.20 Mechanisms for applying a pure torque

A.4.10.
$$r\sum_{1}^{n}\frac{1}{n}:$$

if you could not do it, try a changed viewpoint, starting from the top.

A.4.11. The equivalent kinematic property is that of a mechanism which can prevent rotation without applying any other constraint. Figure 4.20 shows two possibilities.

A.4.12(2). The stresses in any gravity-loaded structure are proportional to ρgL, where ρ is density, g is the acceleration due to gravity and L is a typical length. For the blade g is replaced by $\omega^2 r$

$$= (314)^2 \times 0.6 = 59 \times 10^3 \text{m s}^{-2}$$

so that the equivalent length of steel bridge, ρ being the same, is

$$0.6 \times \frac{59 \times 10^3}{9.81} = 3600 \text{ m}$$

or about three times as long as the longest suspension bridge, which is accordingly the only viable form. This blade was Ljungstrom's experimental suspension blade and its 'sag' took it out to 1600 mm radius, so that it was equivalent to a bridge nearly six miles long. This extreme form gave the large outlet area Ljungstrom sought as a solution to his matching problem within a single machine (section 5.3). He based his design partly on the Golden Gate suspension bridge [22].

5 Matching

5.1 Matching: the windlass

As in many other fields of human enterprise, it is helpful when we come upon an aspect of a design problem that has strong affinities with others we have met. Two such aspects, called here *matching* and *disposition*, will be treated in this chapter and the next.

A rather trivial problem in matching is the design of a windlass for raising buckets of water from a well: Figure 5.1 shows the arrangement, where the

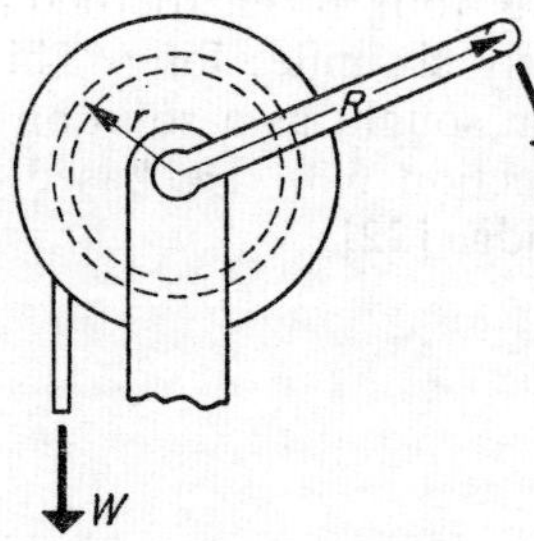

Figure 5.1 Windlass

drum is such that the rope is wound up with a pitch circle radius r. The drum is turned by hand by means of a crank of throw R. The matching problem lies in choosing r and R so as to enable the human prime-mover to put out the maximum power, let us say, or perhaps to raise the full bucket at a given rate with the least apparent effort—various criteria are possible, and ergonomics is scarcely a fine enough tool yet to distinguish between them. Nevertheless we can determine what is a suitable 'throw' for the winder and put R equal to it; we can likewise specify a suitable value of the force experienced at the handle, F—if it is too high, winding will be too hard, and if it is too low, winding will be too slow. Then if we put $r = FR/W$, this particularly easy matching problem is solved.

5.2 An extended example of matching: ship propulsion

Suppose it is required to drive a ship at velocity V by means of steam, and we do this by thrusting some of the surrounding water backwards at an absolute velocity v. Then since the necessary thrust T is exerted on water moving backwards at an average velocity $v/2$, the rate of doing work is

$$T(V + \tfrac{1}{2}v)$$

instead of the TV that would be necessary if T were exerted against a fixed object.

The ratio

$$\eta_F = \frac{V}{V + \frac{1}{2}v}$$

of these two powers is called the Froude efficiency, and it is clearly desirable that η_F should be as large as reasonably possible. We must keep v small relative to V; if to fix ideas we put $V = 10$ m/s and $v = 3$ m/s,

$$\eta_F = \frac{10}{11.5} = 0.87$$

This means we are losing 0.13 of the power used to drive the ship, in the form of kinetic energy of translation in the wake. To reduce this fraction would be desirable, but would generally impose other penalties which would outweigh the gains.

If the steam were expanded through a single reversible nozzle to condenser pressure it might reach a velocity c of, say, 1500 m/s. The ratio $c/v = 500$ gives some indication of the large matching problem that exists in using the small flow of high-energy steam to drive a much greater mass flow of water. A better measure of the water side is provided by the velocity

$$c' = \sqrt{\left\{2\left(V + \frac{v}{2}\right)v\right\}} = 8.3 \text{ m/s}$$

which is the speed the water would have if the specific work done by the propeller were entirely converted into kinetic energy of the water. For the immediate purpose, however, this is not important; all that matters is that the gap to be bridged is a very large one. However, we shall use c' in future instead of v as our typical water speed.

The problem is aggravated by the volume flows involved. The volume flow of sea water is about four hundred times that of the steam in its initial condition, and because of the much lower general levels of velocity in the water the area required to pass it will be much greater than that needed by the steam. It is probable that the radius of the propeller will need to be about 12 times that of the inlet end of the steam turbine (we shall check this later), so that without gearing the steam turbine blades would move only one twelfth as fast as the propeller tips, aggravating the matching problem.

For similar machines we need specific velocities c in the same ratio as the blade speeds. Since c for the steam is 1500 m/s, and the propeller blades move twelve times as fast as the turbine blades, without any form of matching the ideal c' for the water would be $12 \times 1500 = 18\,000$ m/s. The ratio of this speed to the true c', 8.3 m/s, is

$$\frac{18\,000}{8.3} = 2170$$

and this is a rough measure of the gap we have to bridge. 2170 is a sort of gearing ratio, in the widest sense, that we have to provide.

With such a large ratio, three methods of matching are needed:

(a) compounding in the turbine,
(b) toothed reduction gearing, and
(c) different blade forms on propeller and turbine.

Crudely speaking, if we have n pressure-compounded stages in the turbine each will have an enthalpy drop and hence a c^2 of only $1/n$ that for a single stage. The value of c is thus reduced by a factor $1/\sqrt{(n)}$, and so with 20 stages we should achieve a ratio from this source of about 4.5, leaving

$$\frac{2170}{4.5} \simeq 480 \text{ to go}$$

Of course, regardless of the matching problem, there are imperative reasons why the turbine should be compounded, both fluid-dynamical and mechanical, so this much matching costs us nothing. In the early days of turbine ships, however, when attempts were made to avoid toothed gearing, very large numbers of stages were used [23].

A contribution from (c) can be had by making the turbine blades of such a form that they extract the maximum possible energy from the steam relative to their velocity u, i.e. such that the ratio u/c is as low as possible, consistent with efficiency. The form of the propeller blades on the other hand should be chosen so they do little work relative to their velocity u', i.e. so that u'/c' is high. In Figure 5.2, all forms of turbomachinery are shown plotted against two parameters, one of which is u/c. The lowest value of u/c attainable with efficiency (excluding velocity compounding or a stage followed by a diffuser) is a little under 0.5, given by the impulse turbine.

With propellers very high values of u/c can be obtained, but for reasons which will not be discussed here the practical limit for the ship is about 4. Blade form differences, obvious in the contrast between the strong curvature of the turbine blade and the flattened form of the propeller blade, can thus give us a ratio of about 8, leaving a quotient of about 480/8, or about 60 to be taken care of by the toothed gearing.

It must be understood, of course, that this broad and simple study of the ship-propulsion problem is not of any practical use. But it is hoped that it will be of substantial value as an example of the sort of abstract and yet highly pertinent treatment of a matching problem which generates insight and leads to a rapid and quantified grasp of less familiar or less thoroughly studied design situations (see Question 5.2).

The parametric plot (section 2.4) in Figure 5.2 has much in common with the insets of Figures 2.4–2.7 and may also be regarded as a graphical piece of the engineering repertoire (section 2.3). The other axis is of

$$\frac{4Q}{\pi D^2 c} = \psi \text{ (say)} \tag{5.1}$$

where Q is the volume flow, D the diameter, and c has already been explained. This is a measure of the *size of the volume flow* relative to the machine that handles it. To reduce as much as possible the mismatch owing to the diameter

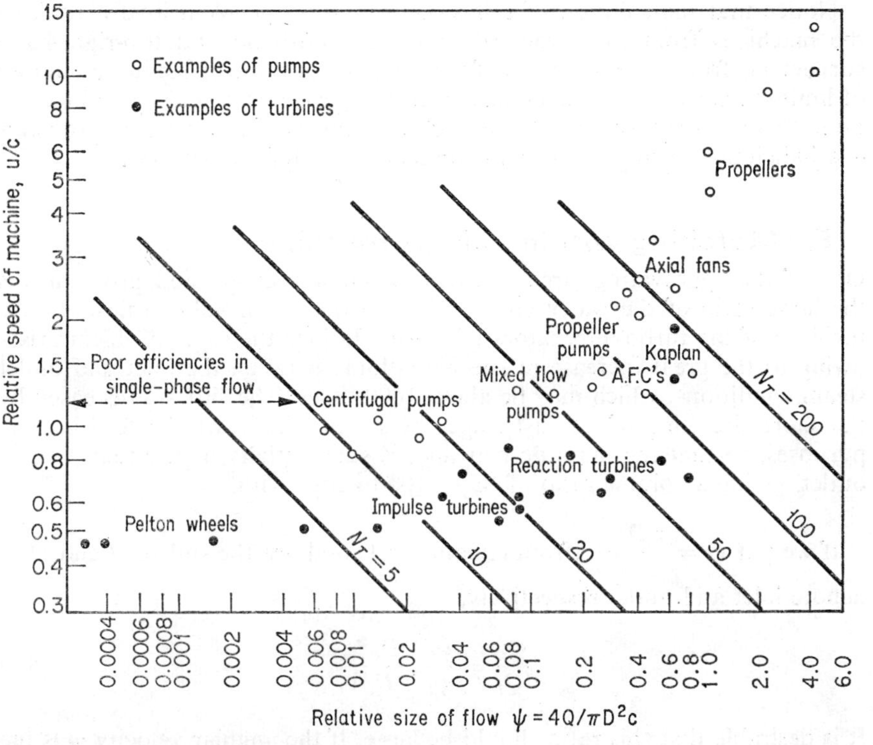

Figure 5.2 Family of turbomachines

of the propeller being much greater than that of the turbine, we must make the propeller as small as possible (i.e., with a high ψ) and the turbine as big as possible (with a low ψ). It can easily be estimated that the propeller will have a ψ' of about 1.2, and we can put, using primes to denote propeller values,

$$\frac{\psi}{\psi'} = \frac{Q}{Q'} \cdot \frac{D'^2}{D^2} \cdot \frac{c'}{c}$$

or

$$\frac{\psi}{1.2} = \frac{1}{400} \times 144 \times \frac{8.3}{330} \text{ (say)}$$

(allowing for the reduction in c by compounding)

$$\text{or} \quad \psi = 0.011 \text{ (say)}$$

Looking at Figure 5.2, it will be seen that this is possible in combination with $u/c = 0.5$, though nearing the limit where efficiency falls rapidly due to over-short blades, windage etc. (Notice that this limit can be exceeded where there is a free surface, as in a Pelton wheel, and the windage is due only to a fluid of much lower density than the working fluid; then high losses are not associated with partial admission.)

Notice that the solution of this ship-propulsion problem involves taking the machines from the (practicable) bottom-left-hand and top-right-hand corners, i.e. from the extremities of the field, and pushing up against a number of limits such as turbine first-stage efficiency, propeller efficiency and cavitation. This is a very common feature of matching as of other design problems —a judicious pressing towards the limits in all helpful directions.

5.3 Matching within a single machine

One of the aggravating circumstances in the ship-propulsion problem was the large ratio of the water volume flow to the steam volume flow. In the turbines of the turbo-alternators of power stations the same problem arises owing to the great increase in specific volume between the inlet and outlet steam conditions, which may be about 1600 times. The situation is eased by the short circuiting of the last stages by steam withdrawn for feed-heating purposes, so that the mass flow at inlet is substantially higher than that at outlet, giving a volume ratio of, say, 1100 to cope with.

If we put $Q = \dfrac{\pi D^2 c\psi}{4}$ (from equation 5.1) and use the suffixes I and O to denote inlet and outlet respectively,

$$\frac{Q_O}{Q_I} = \frac{\psi_O}{\psi_I} \cdot \frac{D_O{}^2 c_O}{D_I{}^2 c_I} \tag{5.2}$$

It is desirable that this ratio should be large. If the angular velocity ω is the same for all stages, i.e. we have only one shaft, then $D = 2u/\omega$, and equation 5.2 can be written

$$\frac{Q_O}{Q_I} = \frac{\psi_O \, (u_O{}^2/c_O{}^2) \, c_O{}^3}{\psi_I \, (u_I{}^2/c_I{}^2) \, c_I{}^3} = 1100 \tag{5.3}$$

Thus again we need to go from a low ψ and a low u/c at inlet towards a high ψ and a high u/c at outlet, sloping upwards to the right in Figure 5.2. We might scrape 40 from the ratio of ψ's, perhaps another 3 from the (u/c)'s and a little from the c's. We shall still be short of our target by the best part of a power of ten.

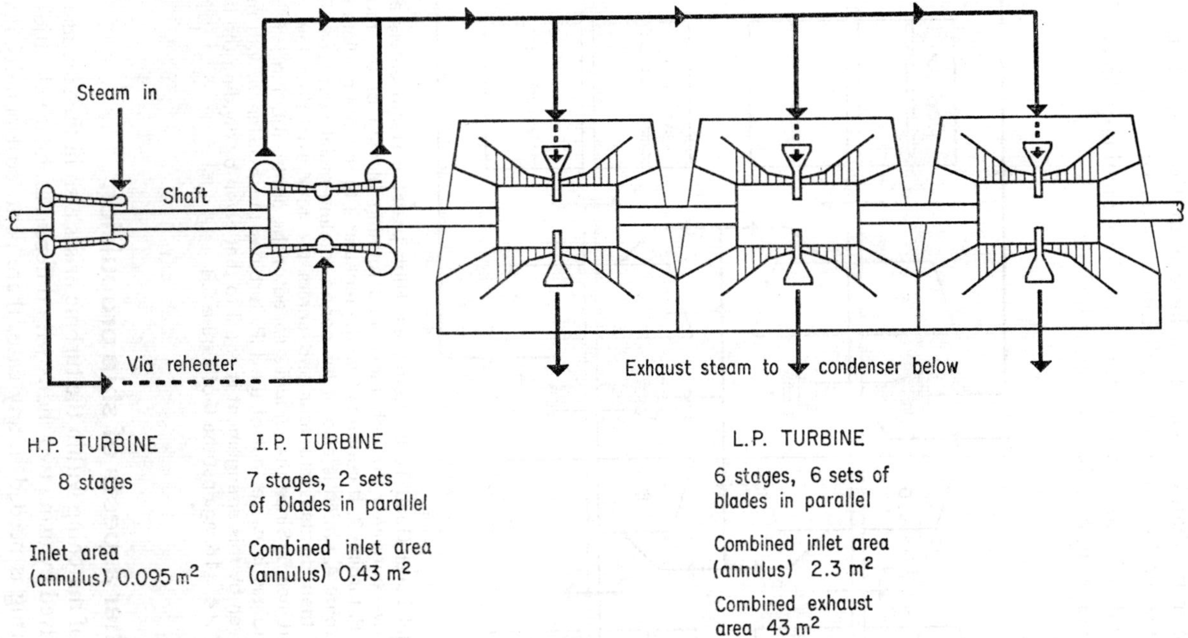

Figure 5.3 Multiple L.P. turbine. (By kind permission of E.E./A.E.I. Turbine-Generators Limited)

The most hopeful ways of bridging the remaining gap are

(a) to have more than one ω, i.e. to have more than one shaft,
(b) to have several turbines in parallel in the later stages.

Figures 5.3 and 5.4 show two solutions to the matching problem between first and last stages in a turbo-alternator turbine.

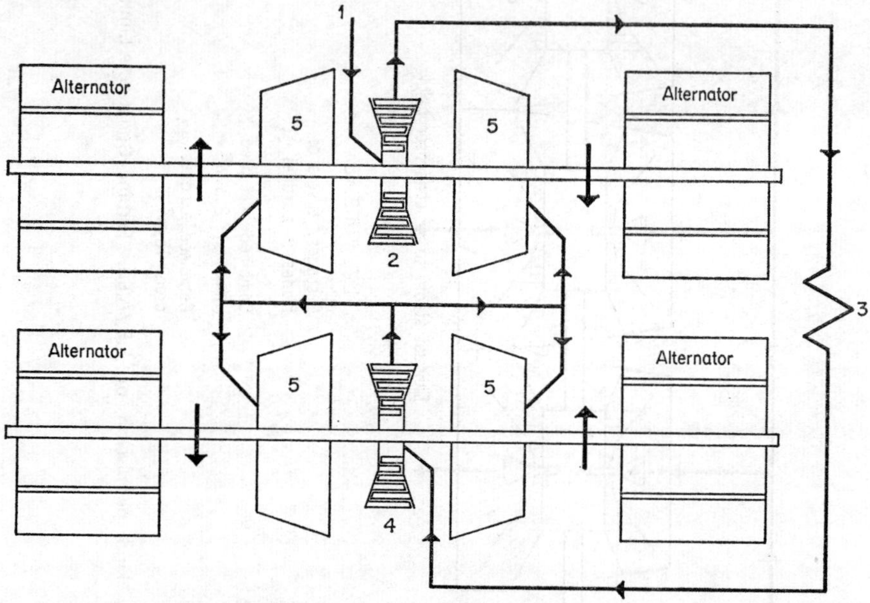

Figure 5.4 Stal-Laval Quad Turboalternators. High-pressure steam enters at 1 and flows outwards through the contra-rotating blading of the radial-flow H.P. turbine 2. It continues via the reheater 3 to the intermediate pressure turbine 4, which is also of the contra-rotating radial-flow type. The flow is then divided into four and enters the four axial-flow low-pressure turbines 5. Since the *effective* speed of the contra-rotating H.P. and I.P. turbines is twice that of the L.P. turbines the assistance with matching given by this arrangement is equal to that given by dividing the flow into $2^2 \times 4 = 16$, against the 6 of Figure 5.3

5.4 Further aspects of ship propulsion

The problem of matching within the turbine arises also in the steamship, and is generally solved by using two shafts with different angular velocities. Since reduction gearing is needed in any case, there is no cost penalty involved; indeed, there is a gain (section 8.2), and the shorter, broader plan obtained by cutting the turbine into two lengths and putting them side by side suits the available space much better.

Before about 1958, the fuel costs of steam turbines and i.c. engines for ship propulsion were about the same, the steamship's boilers burning more but cheaper oil. When it became possible to use boiler fuel in the big slow-running marine i.c. engines, steam was at a critical economic disadvantage. For the largest powers, however, required in the supertankers, advanced steam plants have been designed with fuel consumptions which have been pared down to no more than about 15 per cent higher than the engines. The question has been studied extensively of how to reduce or eliminate this remaining gap.

It is not practicable, by known means, to reduce much further the fuel consumption of the steam plant per shaft kilowatt hour, but if the propulsive efficiency of the propeller could be increased, the same S.H.P. would go further. Now the big marine i.c. engines run at 100–110 rev/min, and this gives too high a value of u/c for propeller efficiency, which would be substantially better at 75–80 rev/min. The i.c. engines could be made to run at this speed, but since their power is roughly proportional to the product of their swept volume and their speed, they would have to be 33 per cent bigger. For the steamship to reduce its propeller speed, however, only requires a somewhat costlier reduction gear (section 8.2). This is an example of a general principle we shall call 'exploitation of assets' (section 8.1); having been forced to adopt this expensive complication, the reduction gear, the designer should not stop short at providing the propeller speed of 100–110 rev/min traditional for the i.c. engine, but press on down, as it were, to the optimum speed, where the rival system is only able to follow at much greater expense. Having accepted gearing, it should be exploited to the full and not simply used to overcome the original difficulty.

However, it is unlikely that the story will end here; already good results are being obtained with boiler fuels in i.c. engines of medium speed (in this context, say, 350 rev/min). Such engines must be geared, and so are in the same position as steam turbines as regards using either lower-speed propellers, or another alternative giving similar improvements in efficiency, contra-rotating propellers (section 8.3). They retain the basic advantage over steam of lower fuel consumption per S.H.P. hour; they have advantages over the big 'cathedral' engines of smaller cost, weight and size, and easier repairs, provision of spares and so forth. If certain problems are overcome, this third contestant may triumph over the other two.

5.5 Specific speeds: degrees of freedom

Some readers who know something of specific speed may be troubled by Figure 5.2, which might be called a Procrustean chart because of the ruthless way in which all turbomachines have been made to fit it. Figure 5.2 is plotted against two parameters; how is it that for water turbines only one parameter, specific speed, will suffice?

For any given requirement for a machine, we could choose u/c and ψ anywhere in the workable region of Figure 5.2: we have two degrees of freedom of choice. In the steamship problem, the requirement of matching turbine and propeller with as little gearing as possible meant choosing them as far apart as possible in value of

$$\psi\left(\frac{u}{c}\right)^2 \quad \text{(see equation 5.3)}$$

which is simply shown to be proportional to the square of the specific speed. Now we have still applied only one condition (namely, a difference of specific speed), but because of the location of the boundaries of the workable region this has dictated our solution completely; in fact, we are reduced to the two extremities of the region perpendicular to the lines of constant specific speed, and we still get only a factor of 8 or so towards our matching requirement of about 2000.

Where there is no dominant matching requirement, as in a water turbine for electricity generation, we have two degrees of freedom of choice (see section 9.2). These must be exercised to reduce the capital cost of the machine and increase its efficiency as much as possible. The general effect of this is to concentrate the economical solutions with no matching requirement along a line, and the position of a point on this line requires only one co-ordinate to specify it, specific speed or ψ, say (u/c would not do because for part of the range the line has constant u/c).

To see a little of the reasons for this concentration of the unmatched solutions in a line (or, rather, a narrow belt), consider a few examples. A medium-head, high-flow turbine could be built as a Pelton wheel, with a value of ψ perhaps 0.01 times that of the reaction turbine which would be the usual choice, and so, since Q and c remain unaltered, 10 times the diameter. Since the reaction turbine might be, say, 8 m in diameter, it is not difficult to see that the capital cost of the Pelton wheel would be excessive. If, on the other hand, we try to invade the home territory of the Pelton wheel with the reaction turbine, the mechanical problems imposed by the high pressure, particularly in the variable nozzles, become incompatible with an efficient solution.

In the range in which the Pelton wheel and the high-head Francis turbine are in competition, and the former suffers from its large size, it should be noted that the $1\frac{1}{2}$ power law holds (see section 4.5), so that if we use three smaller Pelton wheels, each will be of $(1/3)^{\frac{1}{2}} = 0.577$ the diameter and the total weight approximately

$$3 \times \left(\frac{1}{3}\right)^{\frac{3}{2}} = 0.577$$

and this big reduction will be strongly reflected in the capital cost.

In principle it is wrong to optimize the turbine apart from the alternator but the effect of the latter is generally small, for reasons given in section 4.7. The interaction of the turbine speed and alternator cost is not powerful (though sometimes influence arises from questions of attitude; i.e. whether axes are horizontal or vertical).

Many important questions have been begged here, the effects of Mach number and Reynolds number, cavitation, control aspects and so on. Above all perhaps it should be noted that propellers, ducted fans, windmills and some bulb turbines (Röhrturbinen) have in effect an extra degree of freedom since the ratio of the specific work to the kinetic energy or apparent kinetic energy can also be varied. On the other hand, the optimum hub/tip ratio is usually the smallest that can be achieved, so that one of the other degrees of freedom is virtually eliminated, and a two degree of freedom plot is not too 'Procrustean' in its effect.

This discussion of turbomachinery has illustrated, not only matching, but important ideas like exploitation of assets, recognition of limits, division of function among a number of similar units, dimensional methods, and, above all, the generalized way in which it is possible to view this large and very important section of the engineering repertoire.

5.6 Matching of a spring to its task

The windlass and the ship-propulsion are examples of matching by *gearing* in a broad sense, where the application of a constant ratio to an angular velocity or a force is all that is needed. The essential element of *gearing* in a narrow sense is the lever or the pair of toothed wheels, sprockets and chain, pulleys and belts, friction discs etc., which are all embodiments of the lever principle, made continuous in operation.

In this section an example of gearing in a broad sense will be studied, the selection of the form of a spring to suit it to a given task. To start with, consider a helical compression spring which is to exert a maximum force F under a compression in length of a. We saw in section 4.5 that the energy stored per unit volume of the most highly stressed material in a helical spring, working at point C in Figure 3.10, was

$$\frac{(1+\nu)f^2}{4E} \quad \text{(equation 4.9)}$$

and that the efficiency of use η_u was $\frac{1}{2}$.

Thus, ignoring dead turns and stress concentration, the energy that can be stored is

$$\frac{(1+\nu)f^2}{8E}\,v$$

where v is the volume of material in the spring. This must be at least equal to the energy required to be stored in the spring, which is $\frac{1}{2}Fa$. We thus know that the minimum volume of stuff needed is

$$v = \frac{4FaE}{(1+\nu)f^2} \tag{5.4}$$

If we cut off a length L of wire of any radius r within a wide range such that the volume has this value, then it can be made into a spring having the required characteristics. If the coils are flattish, as in a practical spring, the

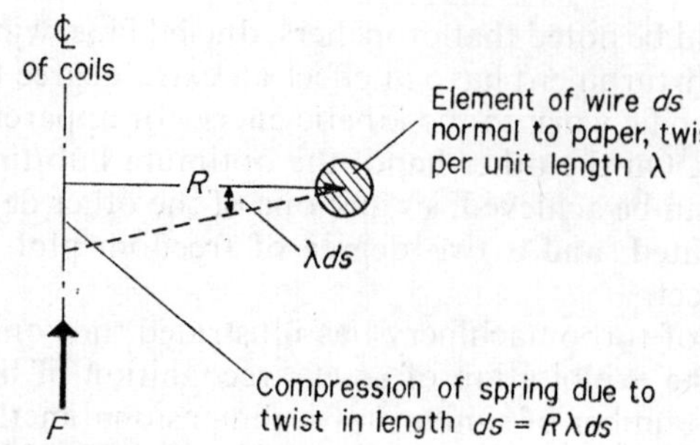

Figure 5.5 Wire of compression spring

wire is nearly in pure torsion, with a constant torque FR in it and a constant twist per unit length, λ, say. The reduction in length of the spring due to the twist in a short length ds of the wire is

$$R\lambda ds \text{ (see Figure 5.5)}$$

so that the total compression is

$$R\lambda L = a \tag{5.5}$$

Now the twist per unit length is given by

$$\lambda = \frac{\tau}{Gr} = \frac{2\tau(1 + \nu)}{Er} \tag{5.6}$$

where G is the shear modulus, τ is the shear stress in the surface, and ν and E have their usual significance in this context (see section 4.5).

If the material is working at point C $(-f/2, f/2, 0)$, $\tau = f/2$ so that from equations 5.5 and 5.6,

$$R\lambda L = R\frac{f(1 + \nu)}{Er} L = a$$

or

$$R = \frac{Era}{f(1 + \nu)L} \tag{5.7}$$

Since L is given when r is known and v fixed by equation 5.4, by the relation

$$v = \pi r^2 L \tag{5.8}$$

equation 5.7 gives a value of R corresponding to each r which suitably 'gears' the wire material to its task. If the wire is very thin, then R will be very small and a design impossible; if the wire is too thick, then R will be too large relative to L to make a satisfactory spring.

Notice that we could have derived an alternative relationship for R by considering the torque in the wire, FR, and relating it to the stress τ by

$$FR = \frac{\tau J}{r}$$

where J is the polar second moment of area of the wire cross-section, which would lead to the same result. In just the same way, if we have a 1000 kW machine driven by a 1000 kW prime mover, we can choose the gearing between to equate either torques or speeds, and then the remaining quantities, speeds or torques, will automatically be equal. The energy storage capacity

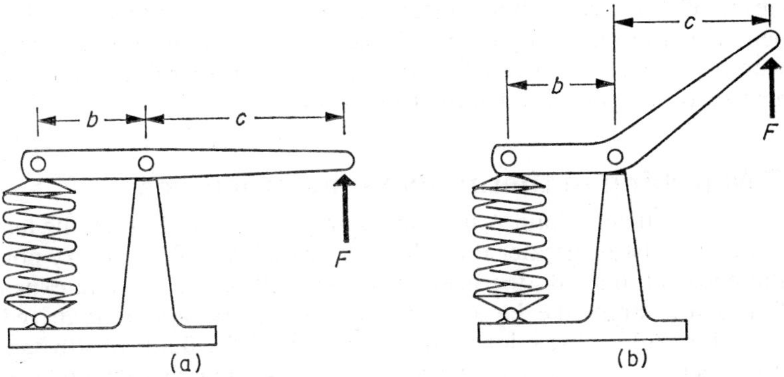

Figure 5.6 Spring 'geared' by lever
(a) point matching (b) curve matching

of the spring fixes the quantity of material, and then if either the force or the deflection is matched, the other requirement is automatically fulfilled.

As an alternative, we could use any spring of the right energy storage capacity and a lever to match it, as in Figure 5.6(a). Suppose P is the full load and d the full-load deflection of the spring to be 'geared' by means of the

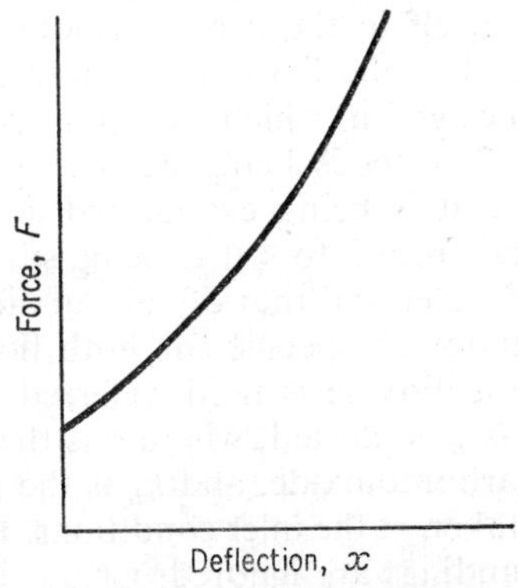

Figure 5.7 Exponential characteristic of suspension spring

lever, then $Pd = Fa$ by the equal energy storage requirement, and from the figure, $Pb = Fc$, so that $d/b = a/c$, a result that could have been found equally well by considering the travels a and d.

By using a nonlinear lever, such as the cranked one of Figure 5.6(b), we can make a linear spring of constant stiffness into a nonlinear one: in Figure

5.6(b), as the deflection increases, the arm c decreases, the 'gearing' from F to the spring increases and so the stiffness at F grows larger. This is useful in such applications as car springs, where it is desirable that the stiffness of the suspension should increase with load in order to keep the natural frequency of vibration about constant.

A spring of this sort, with a constant ratio of stiffness to load, will have an exponential variation of force with deflection, as in Figure 5.7. This is the first example we have used yet of matching a *shape*, rather than a point value. The next section deals with a similar problem.

5.7 Matching in thermodynamic processes

If in a heat exchanger a fluid A is giving up heat to a fluid B, then the temperature T_A of A must be greater than the temperature T_B of B at every point. If the difference is too small, the heat exchanger will be too big and expensive, but if it is too large, the large irreversibility will be apparent in reduced efficiency. This subject was dealt with in section 3.5 for fluids of substantially constant specific heat at constant pressure (c_p), but matching problems arise when the specific heat of one or both fluids varies extremely, as when a pure substance is being heated to its boiling point, evaporated and then superheated in a boiler.

Figure 5.8 shows heating and cooling curves for the heat exchange in the boiler of an early gas-cooled-reactor (G.C.R.) nuclear power station. Along the x axis is plotted specific enthalpy, and along the y axis absolute temperature to a nonlinear scale which is explained later. Curve A shows the variation of enthalpy with temperature of the carbon dioxide which is evaporating the water, cooling itself in the process (indicated by the arrow head from right to left). Curve B_1 (solid line) shows the variation of enthalpy with temperature in a first design in which steam is raised at a pressure of 2.1 N mm^{-2}; from 1 to 2 water is being heated at a constant pressure of 2.1 N mm^{-2}, from 2 to 3 it is being evaporated at the same pressure and constant temperature, and from 3 to 4 it is being superheated.

Since the mass flow of water and that of carbon dioxide are not the same, the enthalpies shown cannot be specific for both flows. For A the specific enthalpy h_A of the carbon dioxide is used, referred to a zero at the outlet conditions, while for B_1 bh_B is plotted, where b is the ratio of the mass flow of water to that of the carbon dioxide, and h_B is the specific enthalpy of the water. The zero for h_B is taken at the inlet conditions. Pressure drops and heat exchange with the surroundings are ignored.

By the steady flow energy equation (first law), the decrease in h_A from 7 to 8 (see the figure) is equal to the increase in bh_B from 1 to 4, and since we have put 8 directly over 1 in the figure, then 7 is directly over 4. Any two adjacent points Q, R, one on the carbon dioxide and one on the steam side, lie directly over one another in the figure, by applying the same argument to the part of the heat exchanger, from QR to either end. Then the condition that T_A is always greater than T_B is that curve A always lies wholly above curve B_1.

Now the scale of T has been chosen so that the total area under the A curve, right down to the heat rejection temperature T_0, gives the available energy lost by one kilogram of carbon dioxide, and the area under B_1 represents the available energy gained by the corresponding b kg of steam (see section 4.5).

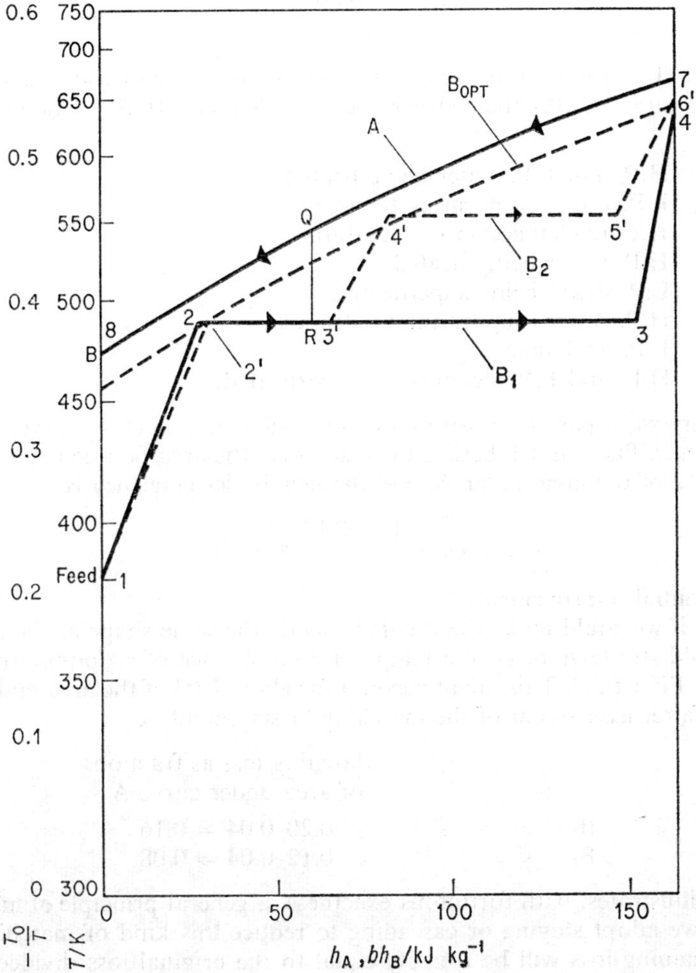

Figure 5.8 Heating curves, dual-pressure boiler

The area between A and B_1 is the lost work per kg of carbon dioxide resulting from the temperature difference in the heat exchanger, and owing to the poor matching of the curves A and B_1 it is very large, about 0.20 of the area under A. Another way of putting this is

$$\text{boiler thermodynamic efficiency } \eta_1 = \frac{\text{increase of available energy of water}}{\text{decrease of available energy of carbon dioxide}}$$

$$= \frac{1 - 0.20}{1} = 0.80$$

Curve B_2 is the heating curve for two separate flows of water, a small one at 2.1 N mm^{-2} (L.P.) and a larger one at 6.5 N mm^{-2} (H.P.). The parts of B_2 are

1′–2′ H.P. and L.P. water being heated
2′–3′ L.P. water evaporating to steam
H.P. unchanged in temperature
3′–4′ H.P. water being heated
L.P. steam being superheated
4′–5′ H.P. water evaporating to steam
L.P. unchanged
5′–6′ H.P. and L.P. steam being superheated.

By using two separate pressures and so producing a heating curve with two steps which fits A much better, the lost work, the area between A and B_2, is only 0.12 of the area under A, and the new boiler efficiency is

$$\eta_2 = \frac{1 - 0.12}{1} = 0.88$$

a substantial improvement.

Even if we could make the B curve exactly the same shape as the A curve, we would still have to keep a temperature difference of economic size, as in B_{OPT} in Figure 5.8. This might account for about 0.04 of the area under A, so that a fairer assessment of the matching losses would be

	Matching loss as fraction of area under curve A
B_1	0.20–0.04 = 0.16
B_2	0.12–0.04 = 0.08

This illustrates, with fortuitous exactness, a general principle of matching, that if we adopt staging or cascading to reduce this kind of matching loss, the remaining loss will be roughly equal to the original loss divided by the number of stages N. Thus going to the dual pressure from the single pressure steam system we reduce the matching loss from 16 per cent to 8 per cent, and if we went to three pressures we might expect a reduction to 5.3 per cent, a further gain of only 2.7 per cent. This same behaviour is exhibited by the joint efficiency of fir-tree roots and some riveted joints, and by other examples of thermodynamic matching such as regenerative feed heating, reheating in turbines and pressure staging in refrigerators.

With increased reactor temperatures the advantage of dual pressure cycles has disappeared. The whole of curve A is shifted much higher.

The importance of matching in thermodynamic design is paramount. Whether we wish to produce electricity via steam, to liquefy natural gas for transport by sea, to refine crude oil, to effect air conditioning by means of an absorption refrigerator (see section 4.10) or to desalinate sea water, the processes we first conceive can be systematically improved in performance by looking for imperfections of matching and reducing them. To heat water from condenser temperature with high-temperature furnace gases is uneconomical because of the large temperature mismatch—we use steam tapped

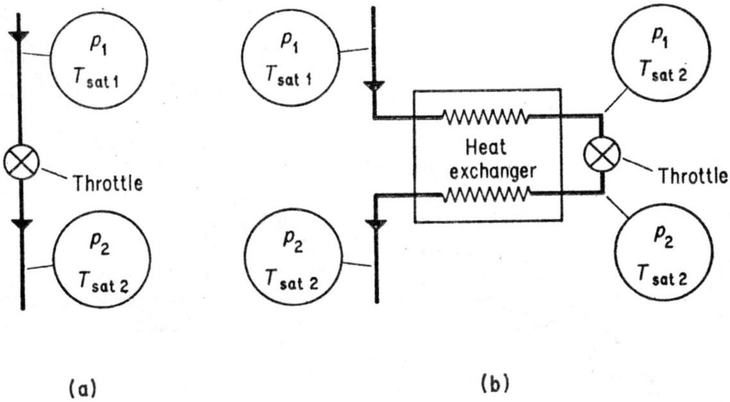

Figure 5.9 Equivalent of isenthalpic throttle
(a) Isenthalpic throttle (b) Equivalent for pure substances

from the turbine instead, at several intermediate stages so as to match the smooth heating curve of the water with a series of steps in the combined cooling curve of condensing steam. The throttling of a liquid refrigerant in stages, the vapour formed at each stage being returned to an intermediate inlet of the compressor, is the functional reverse of this, with the steps in the heating instead of the cooling curve.

More strictly speaking, we improve thermodynamic processes in efficiency by the systematic removal of irreversibilities, and irreversibilities are due to either friction or mismatching, usually of temperature but sometimes of pressure or composition of mixing fluids. Since the frictional irreversibilities are not usually tractable we are left with mismatching. To see why the throttling of a saturated liquid is irreversible mainly due to temperature mismatch, consider the equivalent device shown in Figure 5.9(b) in which the liquid is cooled *before* throttling to the saturated temperature $T_{sat\ 2}$ obtaining *after* the throttle, the cooling being effected by the products of throttling: it will be found that most of the irreversibility in this completely equivalent device is due to temperature mismatch in the heat exchanger.

Figure 5.10 shows the cooling curve of natural gas being liquefied in a very advanced piece of matching. The refrigerant is a mixture, which accounts for the curious form of its heating curve. It is interesting to note that the matching in this process involves composition as well as temperature, and that the 'equivalent device' of Figure 5.9 is no longer equivalent to the throttle but much superior to it when mixtures of substances are being handled, so much so that designs not using it would not be viable.

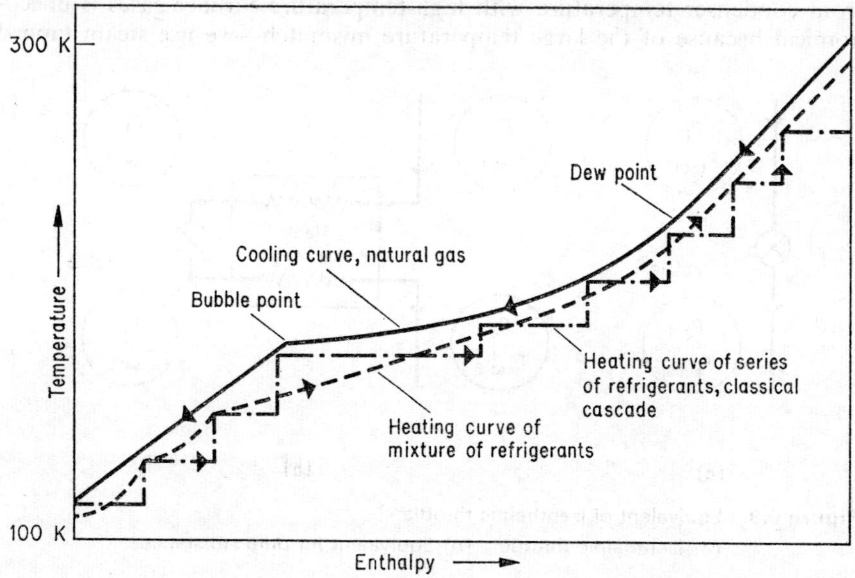

Figure 5.10 Liquefaction of natural gas against a mixture of refrigerants, showing the superiority of matching over that of a more complicated cascade of pure refrigerants. (By kind permission of Technip, Paris)

Matching in all fields should be taken only so far as is economically justified. This will be much further in refrigeration than in prime movers because as a rule only fuel and boiler costs are being saved in the latter, but in the former it may well be fuel, boiler, turbine, compressor, and heat-exchanger costs. Also the shape of the temperature function $1 - (T_0/T)$ in Figure 5.8 means that temperature mismatches of a given number of degrees are more important at lower temperatures.

5.8 Two old cases of matching

(a) Bows

There is some suggestion, from the Mongol victories for example, that the reflex bow of the ancients and the East was more powerful than the longbow

[24]. The conspicuous feature of the reflex bow is its highly prestressed state when strung (see Figure 5.11(a) and reference 25). Figure 5.11(b) shows two hypothetical curves of force (F) against draw (x) of bows. Each has the same force (or 'weight') when fully drawn, and the same draw, so they are in effect matched to the same archer (having a fixed 'corner work'—*cf.* corner power, section 4.6). The heavily prestressed reflex bow will have a sharper initial rise of F and a smaller stiffness later, giving a curve more like R than L, and with more area under it. As the area under the curve is the stored energy, the 'power' of the reflex bow should be greater for the same weight and draw.

It may be objected that the heavy prestressing of the reflex bow means that only a fraction of the strain energy of the bow material is available to drive the arrow, and this is true. As the spring material used had a very high strain

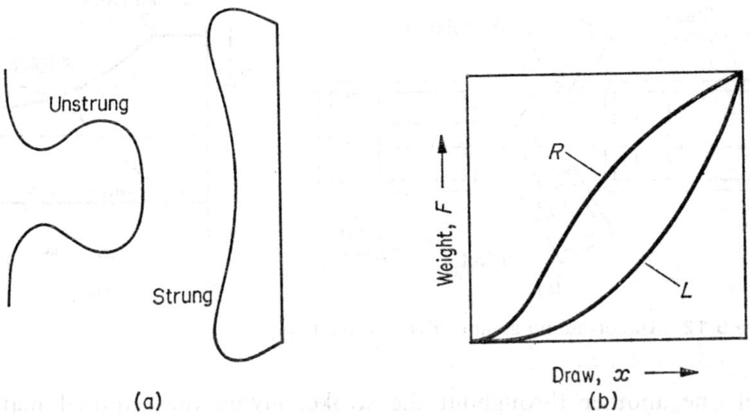

Figure 5.11 Bows
(a) Reflex bow (b) Characteristics (hypothetical)

energy capacity, however, the bow was not heavy: the construction was laminated, with sinew on the tension side, wood in the middle and horn on the compression side—an excellent though primitive example of composite construction to secure the best available material properties at every point for the condition obtaining there [24]. The best animal protein fibres will store about 17 times as much energy per unit mass as spring steel.

An interesting point about all bows is that the ratio of the velocity of the arrow to the velocity of the tips of the bow increases rapidly as the string nears the bow. Kinetic energy which is stored in the bow material immediately after release is mostly imparted to the arrow via the string, so that very little kinetic energy is left in the bow afterwards.

(b) Reciprocating direct steam-driven water pumps

Early in the 20th century, direct-acting steam-driven water pumps were used, of the sort shown diagrammatically in Figure 5.12. The force exerted on the piston assembly by the steam is uniform up to the cut-off C, and then decreases as the steam expands (see Figure 5.12). The resisting force of the water will be constant, provided the back pressure on the pump is uniform. If the average steam force is equal roughly to the water force, as must clearly be the case, there will be an excess of steam force at the beginning of the stroke and a deficiency at the end. In a high-speed machine with a crankshaft, this mismatch would be smoothed out by a slight fluctuation in the kinetic energy of the flywheel. In the slow-speed pumps in question, however, it was desirable to match this unbalanced force by the two air cylinders shown in the figure, which acted simply as springs. The three forces on the piston assembly, all plotted as positive when acting to the right, can be seen to nearly

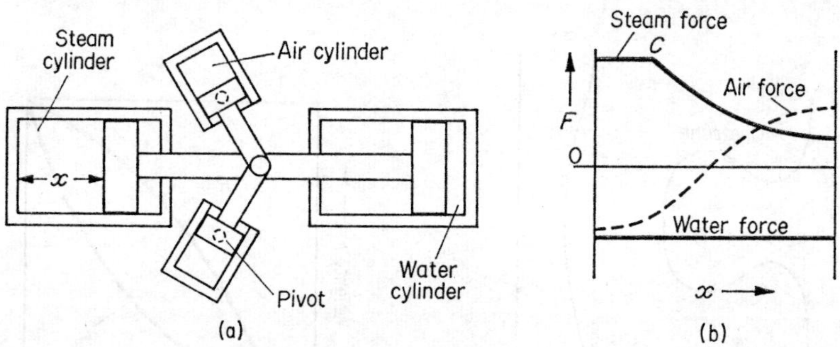

Figure 5.12 Direct-acting steam-driven water-pump

cancel one another throughout the stroke, giving the required matching between the steam and water forces.

5.9 Car handbrake

Consider the problem of incorporating a handbrake in a motor-car disc brake by means of a manually operated mechanism acting on the same brake pads as the footbrake. This, like the design of a bow, is a corner-work problem (see section 4.6).

Figure 5.13(a) shows the principal parts concerned, the disc itself and the two opposed pads of friction material which are clamped onto the disc by nipping forces F when the handbrake is applied. When the footbrake is applied the pads are also clamped, but we can largely ignore this aspect for the purposes of this example, though not, of course, in the design.

It is required that the force F shall have a certain value (F_N) when the handbrake is fully applied in order that it shall be capable of holding the car on some specified gradient, say, 1 in 4. It is also necessary that the clamping mechanism shall have sufficient travel d to take up clearance between the pads and the disc and elastic deflection under F_N in the caliper or other structure closing the force path (shown dotted in Figure 5.13(a)). Note: the drawing of such force paths is often helpful in visualization [44]. While the elastic deflection under F_N is constant, the clearance or backlash between the pads and the disc is variable, so that upon applying the brake the force/travel curve may take either of the forms 1 and 2 in Figure 5.13(b), or any similar shape in between. The flat part at the beginning of each curve is the portion of the travel in which clearances are taken up, and after that is a triangular piece in which further travel is solely due to the elastic deflection of the structural parts.

To fix ideas, let us assume that the clearance between the pads and the disc before applying the handbrake can range from $a/2$ to a on each side, and

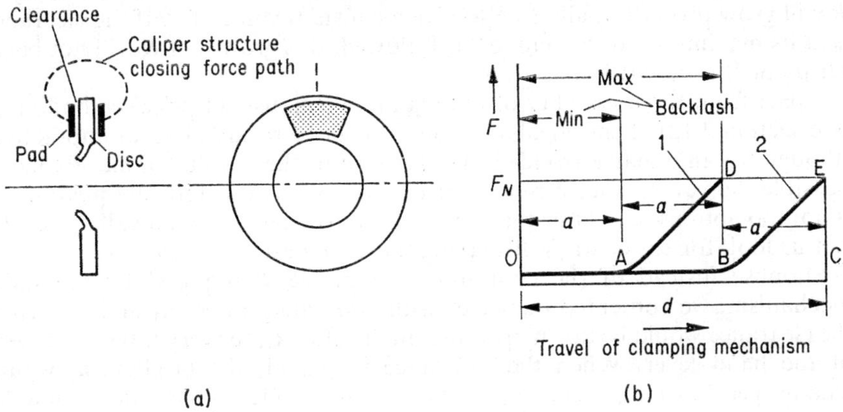

Figure 5.13 Disc handbrake

there is no other backlash in the system. Let us further assume that there is no friction or elasticity in the system except the elasticity in the caliper, which deflects a under the load F_N. Then the actual work done by the driver is the same as that done at the clamping mechanism, and Figure 5.13(b) has the proportions shown, where OA = AB = BC = a. Whether for curve 1 or curve 2, the actual work done by the driver per brake is $\frac{1}{2}aF_N$.

Suppose there is a fixed mechanical advantage R between the handbrake lever and the clamping mechanism, so that the force per brake P exerted by the driver must be F/R at all times, with a maximum value $P_N = F_N/R$, then the travel of the handbrake lever in the worst case (curve 2) is $3aR$ and the *corner work* (section 5.8) demanded of the driver, defined by (maximum force × travel), is

$$3aR \times \frac{F_N}{R} = 3aF_N \text{ per brake.}$$

This is so large in practice as to be an embarrassment in a real system with friction, elasticity and backlash in the other parts. Observe that corner work, like real work, is not changed by gearing of a fixed ratio R, but unlike real work, it is changed by a variable R.

An 'ideal' mechanism is theoretically possible in which the handbrake lever would always travel its full distance at a uniform value of P. Here the force/travel relationship at the driver's end is rectangular and the corner work is therefore equal to the actual work or $\frac{1}{2}aF_N$, a reduction to one-sixth which would more than overcome the problem of 'heavy' handbrakes. This ideal mechanism would be perfectly matched to the load, whether it followed curve 1 or curve 2 of Figure 5.13(b) or any intermediate shape.

Suppose the pad clearance to be at its minimum, so that the resistance at the brake follows curve 1 in Figure 5.13(b). Then while the clamping mechanism travels from O to A, R will be very low in the ideal device, and the hand lever will move only a small distance. As resistance is encountered at A, R will grow proportionally to F to give a constant value of P. If the clearance is at its maximum, so that curve 2 is followed, this growth of R will not begin till point B is reached.

Apart from the possibility of finding an ingenious 'cut price' embodiment (see section 2.12) of this ideal solution, it must be regarded as an impractical (though not impracticable) ideal. It does give us the absolute limit of what is possible, however; since a reduction to one sixth is theoretically possible, it is not too optimistic to hope that a reduction to one half is practically possible. Let us look for easier ways of making more modest improvements.

Suppose instead of the continuously variable R required by the ideal mechanism, we contented ourselves with two values of R, R_1 and R_2. Then the clearance would be taken up at the small value R_1, for very little movement of the hand lever. When the resistance increased, the mechanism would 'change gear' to a high value R_2 at which the remainder of the action would be completed. If we ignore the little bit at R_1 the travel of the hand lever would be aR_2 and the maximum value of P, F_N/R_2. Thus the 'corner work' would be aF_N, an improvement of three times over the fixed R mechanism.

This is still a rather advanced mechanism. The ideal version matches itself perfectly to whatever force/travel relationship arises, curves 1, 2, or in between. This second version matches itself within the limits of its two 'gears,' R_1 and R_2; it still requires some device which reacts to a certain level of resistance by changing gear. Figure 5.14 shows a hypothetical form of such a mechanism.

A smooth bore in a block G houses a sliding plug H and block L, the latter acting directly on a pad and thus being subject to the resistance F. The plug H is hollowed out to a thin-walled cylinder at the end adjacent to L and a rubber cylinder K is bonded in: the other end contains a small sliding piston J, which is pushed by the hand lever via a mechanism of fixed mechanical advantage R_1.

When F is small or zero H and L slide freely in the bore and $R_1P = F$. As F, and hence P, increases, however, the squashed rubber expands the skirt of H, which locks in the bore. The rubber then behaves rather as the fluid in a hydraulic ram, with J the loading piston and L the ram, so that the force R_1P is magnified roughly in the ratio of the cross-sectional area of J to that of K.

This 'cut-price' solution is still rather elaborate, however. Let us remove altogether the need for behaviour sensitive to resistance by using a mechanism in which R varies in a fixed way with travel, matched to curve 1 as far as point B and keeping R unchanged thereafter, i.e. a mechanism perfectly matched to the imaginary load curve OADE, which is the envelope of all possible load curves. It is clear that this arrangement provides more mechanical advantage than is necessary for all load curves except 1.

Since this last mechanism is perfectly matched to the load curve OADE, its corner work is equal to the actual work of OADE, or $\frac{3}{2}aF_N$. Thus a very useful reduction to one-half has been achieved with a relatively simple approach. The shape OADE is not easy to match, and one existing design which

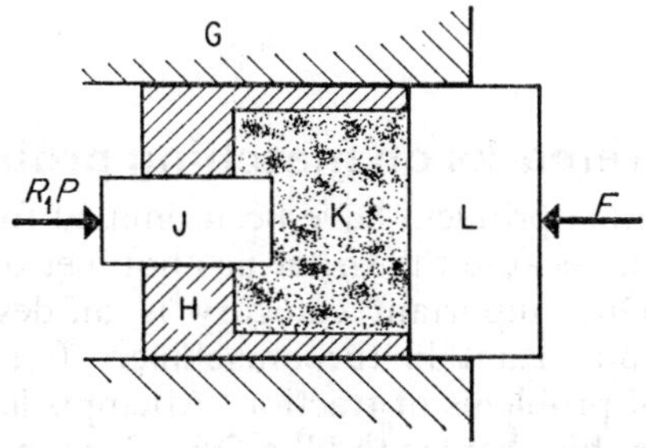

Figure 5.14

uses this kind of matching has a cranked lever, similar to that of Figure 5.6(b) in principle. However the problem has been over-simplified in this account and a smoother matching curve may well be desirable when other effects are taken into account. The cranked lever in question is also a good example of exploitation of assets (see section 8.1), since it serves another purpose.

We have tried three levels of matching to reduce the 'corner work' demand of the handbrake—the best possible matching, a simpler version which still adapts itself to the actual variations of load curve, and the simplest, which takes no account of variations in load curve; the reductions effected are to one-sixth, one-third, and one-half, respectively. The first two approaches might be called 'adaptive' and are rather elegant but difficult to embody in a practical form.

Before leaving this example, another way of improving the handbrake matching should be considered: we might, as in the steam-driven pump of section 5.8, store energy in the first part of the hand-lever travel to be released in a way helping brake application in the second part. If we used a spring, we would merely be adding to the force/travel curves of Figure 5.13(b) a fixed curve of zero area, like the air-force curve of Figure 5.12.

On the other hand, the energy might be stored in a flywheel, which behaves rather like an 'adaptive' mechanism in that it generates torque to overcome any resistance so long as it continues to have kinetic energy left in it. The flywheel may be regarded, in fact, as a simple and perfect adaptive device in this sort of context; while unsuitable for several reasons for the brake, it is of course used in the same matching role on the starter of a car, the starter motor armature providing the flywheel.

We have emerged from this study of the matching problem of a handbrake without any very good solutions, the best being the fixed variation of R with travel. For further improvements attacks might be made on the caliper flexibility and also, an aspect which has been left out but is in practice most important, the flexibility of the transmission between hand lever and brake (see Question 5.11).

5.10 General remarks on matching problems

Many types of matching problem have been omitted for the sake of brevity and others, no doubt, because the writer has not met them or has forgotten to consider them. One important category is off-design-point matching, which is especially important in turbomachinery (see Question 5.8), and another is the special problems of traction. Attempts have been made in the questions that follow this chapter to fill a few of the more conspicuous gaps and to show a little more of the range of guises in which a need for matching may present itself and the means that can be adopted to meet it.

Questions

Q.5.1(1). An electrical power source may be regarded as a generator of voltage V in series with a resistance R_1. This source is to be used to power a device that behaves as a resistance R: what value must R have so that the power expended in the device is a maximum?

Q.5.2(2). A cryogenic plant is to have a turbine that expands a flow of helium from 40K and 0.38 N mm^{-2} absolute to 26K and 0.11 N mm^{-2} absolute. It is proposed to use the power from this turbine to compress the same flow of helium from 300K and 0.11 N mm^{-2} absolute to as high a pressure as possible, by means of a centrifugal compressor. Can the turbine and compressor be matched without gearing?

Q.5.3(3). Figure 5.15 shows a simple letter balance, in which the only weights that need be considered are those of the letter, W and the balance weight, w. The extremes to be weighed are 1 and 16 units, and the relative sensitivity (rate of change of θ with fractional change in W, i.e. $W\frac{d\theta}{dW}$) is to be the same

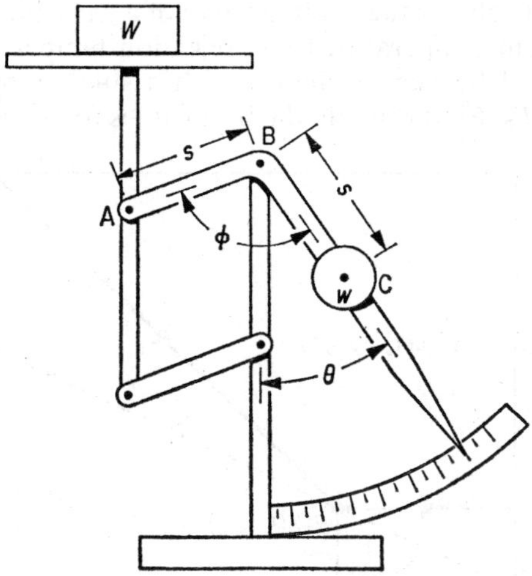

Figure 5.15 Letter balance

at each of these values of W. Find values of w and ϕ that fulfil these conditions and make the minimum sensitivity as high as possible.

Q.5.4(2). Figure 5.16 shows the mechanism of a level-luffing crane. The jib OA makes an angle θ with the horizontal and is 14 metres long. The jib-head

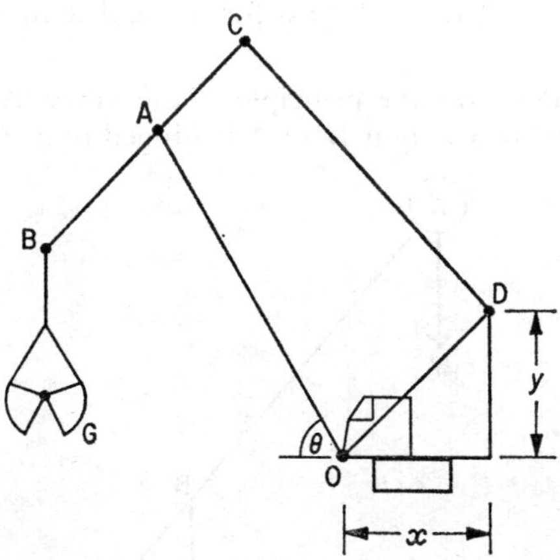

Figure 5.16 Level-luffing crane

BC is 9.8 metres long and BA is 7 metres. The hoisting wire passes round small pulleys at B, C, and D to the cab. D is fixed relative to the cab and CD is a tie of fixed length. Find a position for D such that as θ is varied without the hoisting wire being reeled in or out, the grab G remains at a nearly constant level ('level luffing'), given that, when θ is 75°, BC makes an angle of 15° with the vertical. (Make the level of G the same for $\theta = 45°, 60°, 75°$.)

Q.5.5(3). Figure 5.17 shows the cooling curve of a gas which is being liquefied. The effective ambient temperature (heat-rejection temperature) is 300K. The gas is to be liquefied by heat exchange with a single stream of perfect gas heating from T_1 to T_2. Find roughly the best values for T_1 and T_2, other things

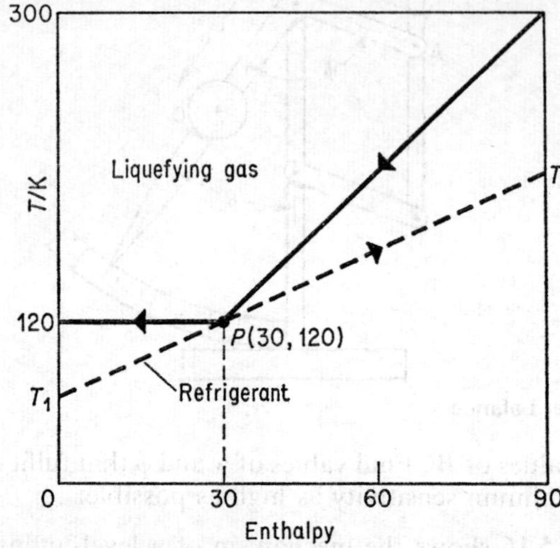

Figure 5.17 Cooling curve for gas liquefier

being equal; ignore the practical need for a small temperature difference at the point P.

Q.5.6(4). Figure 5.18 shows the principle of the Terry 'Anglepoise' lamp. A light arm ABC carrying a weight W at A is hinged to a vertical rod BD at B.

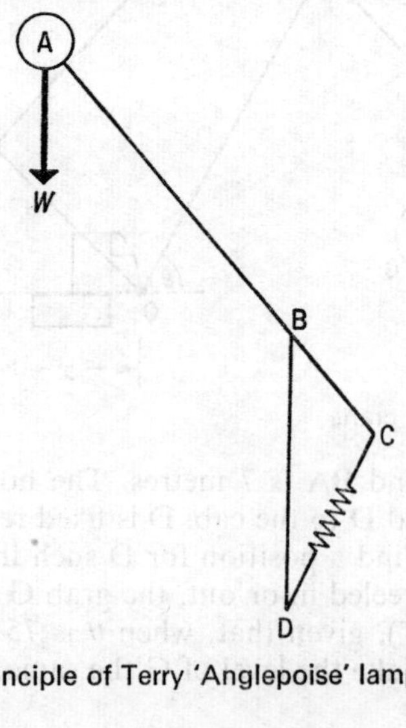

Figure 5.18 Matching principle of Terry 'Anglepoise' lamp

The linear spring CD is chosen so that it just counterbalances W in all positions, i.e. the arm ABC is in neutral equilibrium. Find the unstretched length of the spring.

Q.5.7(?). Figure 5.19 shows in plan and aft elevation a typical arrangement of the lead of the mainsheet on an International 10 m² sailing canoe, a fast

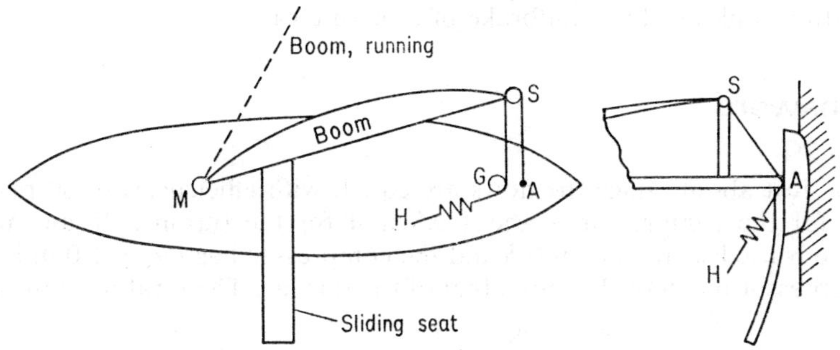

Figure 5.19 Sailing canoe mainsheet arrangement

single-handed craft. The sheet, which serves to haul in the mainsail attached to the boom, runs from a fixing A through a single block S attached to the boom and back through a guide block G to the helmsman at H, who thus has a constant mechanical advantage of two. The ideal system has a variable mechanical advantage, ranging from less than one, when the boom is right out (dotted line in plan) and the force required at S is small, to about 3 with the sail hauled in hard, when the force required at S is greatest. Devise an arrangement better than that of Figure 5.19 in which the boom can be hauled in faster to begin with and there is less rope lying around when closehauled. Note that the large downward component of the force at S when the boom is hard in is desirable as it pulls the sail into a flatter form. The writer has failed to solve this problem.

Q.5.8(3). Consider the problem of matching the circulating compressors of a gas-cooled reactor at part load. These compressors pump the circulating coolant gases through the reactor and boilers, a load which behaves roughly as a simple orifice, i.e. the pressure drop is proportional to the square of the mass flow: it would be easy to match the compressors at part load by dropping their speed, but the cheapest drive is by synchronous electric motors.

Q.5.9(3). Consider the problem of devising a safe-loading indicator for a vehicle-mounted crane where the safe load varies not only with the angle of the jib to the horizontal, but also the angle it is slewed to relative to the axis of the vehicle. Assume that stability is always more critical than strength.

Q.5.10. In a thick-walled homogeneous cylinder subject to internal pressure, the total circumferential stretch at the inside is slightly greater than that at the outside, since the radial stresses are compressive. Thus the product of the circumferential strain and the radius falls with radius, and so the hoop stress falls rapidly with radius: this means that, if the thickness is comparable with the inner radius, the outer layers of material are relatively ineffectual as long as no yielding occurs. This is a serious problem in the design of pressure vessels for very high pressures. List the means that might be employed to match the stresses better.

Q.5.11. Discuss the choice of gear ratio between the hand of the driver and operating cables of the handbrake of a private car.

Answers

A.5.1. R_1.

A.5.2. Just about: since the flows are equal, with efficiencies of 80 per cent the c for the compressor is about 0.8 that for the turbine. If an impulse turbine is used with a u/c of 0.5 and the compressor has $u/c = 1.0$, this gives a compressor diameter 1.6 times that of the turbine. The ψ ratio is thus about

$$\frac{300}{40} \times \frac{38}{11} \times \frac{1}{1.6^2} \times 1.25 = 12.7,$$

(see equation 5.2) which is just about possible (*cf.* Figure 5.2). However, in such a plant the turbine efficiency is much more important, since one unit of heat extracted at 30K takes 9 units of work even in a reversible machine, and probably 15 in a real one. Hence we should design the turbine to be as efficient as possible, and then a single centrifugal impeller may not match.

A.5.3. 4 units, 118°. The sensitivity is

$$W.\frac{\mathrm{d}\theta}{\mathrm{d}W} = \frac{\sin(\phi - \theta)\sin\theta}{\sin\phi}$$

which is symmetrical about $\theta = \phi/2$; when AB, BC make equal angles with the vertical the sensitivity is a maximum, hence W must then be the geometric mean of its extremes $= \sqrt{1.16} = 4$. The sensitivity when $W = 1$ or 16 then turns out to be

$$\frac{4\sin\phi}{17 + 8\cos\phi}$$

which is found to be a maximum when $\cos\phi = -\frac{8}{17}$, or $\phi = 118°$.

A.5.4. $x \simeq 4.5$ m, $y \simeq 4.9$ m. The jib and jibhead can be drawn in each of the three positions; the distance GB does not change in luffing so B also has the same level in all three. Thus three positions of C are found, and D is the centre of the circle drawn through them.

By drawing velocity diagrams the slopes of the path of B may be found for each of the three positions, and hence the path itself may be sketched very closely.

It is interesting to observe the degrees of freedom of choice that exist for improving this solution. We can vary the angle BAC and the length AC, and also the angle of the jib at which it and the jib head make equal angles with the vertical. We can lead the hoist round a large pulley at C, we can move the pulley away from C slightly, and we can take the hoist to a point other than D. Altogether about eight degrees of freedom can be mustered besides the two coordinates of D, which might enable us to make B level in eleven positions. This would be too difficult, however; five points ought to be quite good enough.

A.5.5. Approximately, $T_1 = 75$K, $T_2 = 210$K gives the least matching loss. The loss of available energy of the perfect gas

$$-\Delta b = \int_{T_1}^{T_2} T_0 \mathrm{d}s - \mathrm{d}h = \int_{T_1}^{T_2} \mathrm{d}h \left(\frac{T_0}{T} - 1\right)$$

$$= c_p T_0 \ln \frac{T_2}{T_1} - c_p(T_2 - T_1)$$

The mass flow is inversely proportional to $T_2 - T_1$ so we have to minimize

$$\frac{\ln T_2/T_1}{T_2 - T_1} \quad \text{or maximize} \quad \frac{T_2 - T_1}{\ln T_2/T_1}$$

which is a temperature and may be recognized as the 'log mean temperature'. This may be got up to 131K. However, $T_1 = 50$K; $T_2 = 260$K, and $T_1 = 100$K, $T_2 = 160$K involve increases in ideal work of only about 2 per cent.

A.5.6. The unstretched length is zero. Resolve the spring force at C into components parallel to BD and BC. The latter has no moment about B, so the BD component must be constant and equal to $(AB/BC)W$. Regard BCD as a triangle of forces for C in which BD represents the BD component to a certain fixed scale and so CD represents the spring force to the same fixed scale. Therefore the spring force is proportional to CD, and the unstretched length of the spring is zero. The spring is prestressed so that the coils close up tight together when free and need a certain minimum force to start separating them. this force is the stiffness times the free length.

The 'Anglepoise' lamp is a beautiful example of matching. It has two degrees of freedom that require balancing, since another rod is jointed to ABC at A; both are balanced by springs of the type described, both sets of springs being at the base. A parallelogram mechanism carries the balancing force to the upper rod.

A.5.7. No practical answer is known to the writer. The problem would be much easier if the higher mechanical advantage were required with the boom right out: a single stop would then serve where the actual problem requires some sort of active clamp that positively 'takes hold' of another part of the mechanism, what the writer calls an 'active stop'. There is a fundamental law of mechanisms involved [26]. The writer's best suggestion is a windlass with a conical drum at A with a single rope straight to S. The drum is turned by the helmsman pulling a rope wound on another cylindrical drum on the same axle. This axle runs in a fixed coarse pitch nut that shifts it axially to control the feeds on and off the drums. (Why not use a ratchet instead of the cylindrical drum?)

A.5.8. With constant-speed machines the principal matching means available are throttling, by-passes, shutting down some machines out of a number in parallel and variable-angle blading. Speed control can also be had by a fluid coupling with variable filling, which can be outside the shielding.

A.5.9. Two possibilities are:

(a) a straightforward load measuring device with a warning device set by a very simple analogue computer (probably mechanical) fed with angles of luff and slew, jib extension, etc.,

(b) a device using load cells built into the underframe.

Note that (b) is not affected by a sloping terrain as (a) is.

A.5.10. Three possible means are:

(a) prestress, i.e. induce an initial tension in the outer layers balanced by a compression in the inner layers. This method is used in gun barrels,

(b) make the cylinder as two (or more) concentric tubes with a small gap between, and admit a fluid at a suitable intermediate pressure to the gap. This is only practicable where some sure supply of intermediate pressure is available, as in the H.P. cylinder of steam turbines. In this case there is the further advantage that the outer cylinder, which does not have to withstand such high temperatures as the inner one, can be of cheaper, stronger material (*cf.* section 9.2),

(c) make the outer layers of material of higher E (this is not so remote as it sounds).

A.5.11. The chief object must be to keep up the stiffness of the system as 'seen' by the brake pad. A cable may be regarded as a spring of stiffness s, say. If the gear ratio between cable and pad is R, then a force F at the pad produces a force F/R in the cable, which stretches it F/sR. The movement at the pad due to this stretch is $1/R$ times as much, or F/sR^2, i.e. the stiffness of the cable 'seen' by the pad is sR^2, so R should be large. The practical limit is set by the inconvenience of the large cable travels which result from a large R, e.g. in the accommodation of the levers used to balance the loads in the cables. Another consideration is that a large reduction R is difficult to arrange at the brake, and it helps in this respect to put part of the overall gear reduction at the driver's end.

6
Disposition

6.1 Problems of disposition

A problem of disposition exists where a limited quantity of some commodity, usually space, has to be shared out to the best advantage between a number of functions. Cases which meet this strict definition are not very common, but cases exhibiting similar features abound, so that the approaches discussed in this chapter are of very wide application.

Some problems of disposition reduce to formal optimizations like those of Chapter 3. Some can be exactly optimized by simple reasoning. Many others can be brought to a practical optimum by such reasoning, once the problem has been suitably framed.

A historic example of a problem of disposition, the commodity being the centre distance of cylinders in an i.c. engine, is given in reference 27.

6.2 Dovetail fixing

Consider the design of a dovetail fixing for small blades in axial flow compressors (see Figure 6.1). The female form is to be broached axially through

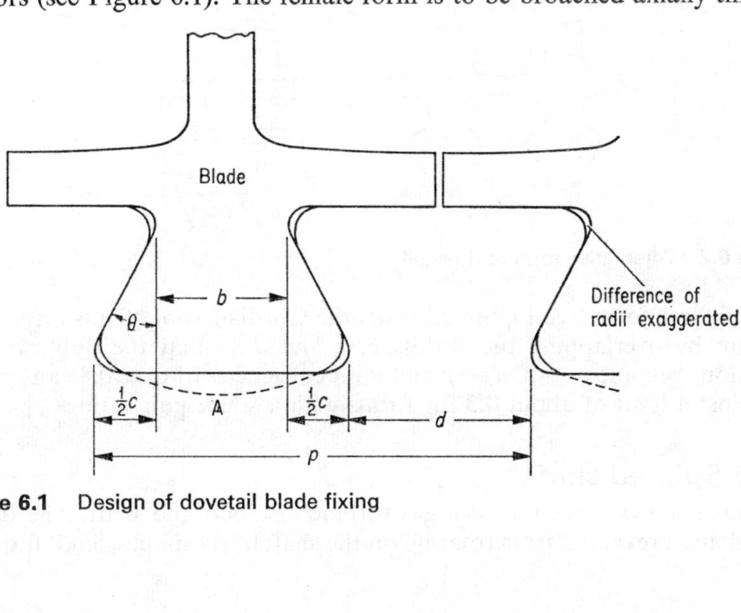

Figure 6.1 Design of dovetail blade fixing

the parallel-sided disc rims and the male form broached or ground on the blade.

For simplicity the materials of blade and disc will be assumed to be metals of similar strength, and the load will be assumed to be direct, without significant bending. The limited commodity is the pitch p, which has to be shared out between three uses, the blade neck width b, the overhanging portions of combined width c, and the disc neck width d, since

$$p = b + c + d$$

Clearly the division of p should be made so that the load at which all three types of failure (blade neck fracture, disc neck fracture, and shearing or crushing of the lugs) are calculated to occur is the same. For the given conditions, this implies making b equal to d, and c about the same, depending slightly on the material properties (for high tensile aluminium alloy, c would be relatively larger than for ferritic steels: the ratio of c to b and d and the angle θ are best determined by experiment).

This treatment is, of course, superficial. The load designed for should include a vibratory bending term, and there are additional centrifugal loads on the disc neck, and, to a lesser extent, the lugs due to the inertia of the fixing itself. Above all, the form is traditional rather than logical; at the very least there should be some reverse batter on the lugs (dotted line, Figure 6.1). The forces in the disc necks have to flow down into the web of the disc, skirting the area A which mostly has the effect of increasing the stress concentration (on the other hand, it does carry hoop stresses which do some holding together: see section 3.11). Also, since the load has to be gathered up axially into the thin web, some thinning of the disc profile should start from immediately below the blade (see Figure 6.2).

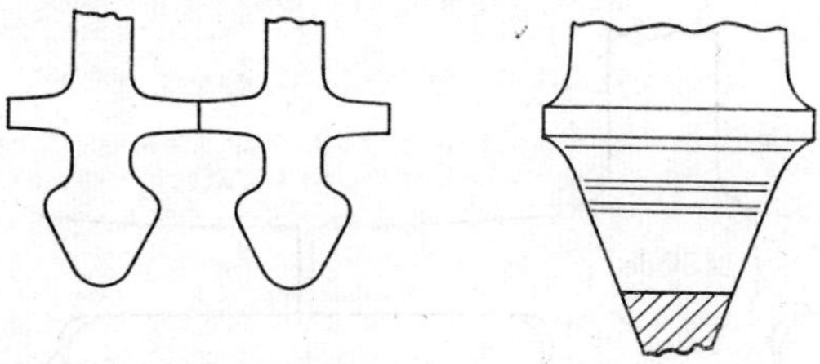

Figure 6.2 Alternative form of dovetail

The fir-tree form (see Figure 2.2) cuts the Gordian knot of this disposition problem by overlapping the widths b, c and d so that the only absolute limitations become $b < p$, $d < p$ and joint efficiencies of over 0.8 can be had, as against a limit of about 0.3 for forms with a single pair of lugs.

6.3 Splined shaft

Figure 6.3 shows a section of a gas turbine disc-hub and shaft. The disc is centred and prevented from rotating on the shaft by six splines, flank fitting so that centring is not affected when centrifugal stresses cause the disc to grow relative to the shaft (by about 0.003 of the diameter). To keep the shaft stiff, it should be as big as possible: to keep the disc weight down, the hole in it should be as small as possible. Thus cross-section is the commodity we have to share out. Consider the area of cross-section lying in the annular region between the radii r_1 and r_0. Clearly, this cannot be given to either disc or shaft as an unbroken ring. But as broken pieces it is of great value on the shaft, since it adds to the second moment of area of cross-section greatly, being at the maximum radius. As broken pieces it is of no value, or even

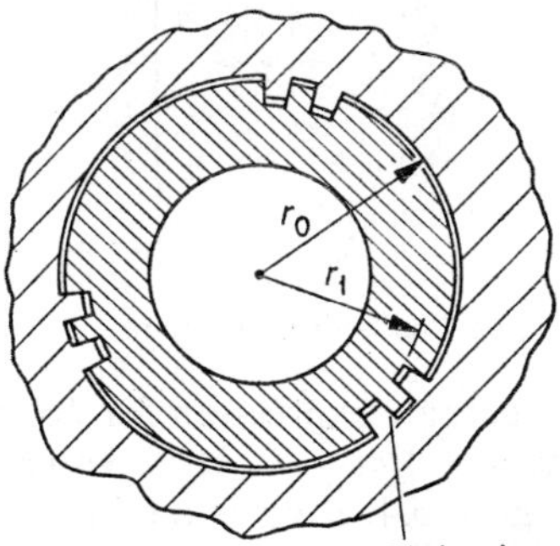

Figure 6.3 Splined shaft and hub design

negative value, on the disc, since it cannot carry any extra hoop stress and long unbroken pieces will cause a greater stress concentration. Thus the right way to design the cross section is as shown, with a minimum number of shallow grooves in the shaft and shallow narrow splines in the disc bore (torsional stresses are low).

This sort of disposition problem can become very elaborate on stepped splined shafts with several sets of cooling air passages. Notice the important idea of attributing different values to particular portions of the cross-section (the commodity), according to the use to which they are put. We then try, as far as possible, to allocate areas to uses so as to maximize the sum of values (see next examples).

6.4 Disc brake caliper: breaking logical chains

Figure 6.4 shows a disc brake in section. The extreme radius over the caliper is limited to a fixed value R by the requirement to clear the wheel. It has already been shown that an important measure in this design problem is the *corner work* $F_N d$, where F_N is the nipping force over the pads and d is the effective maximum travel of F_N, including a, the elastic deflection of the caliper, and the clearances between the pads and the disc.

Now if the radius to the centre of the pads r is reduced, then F must be increased in the same ratio to maintain the same braking torque. The stiffness of the caliper can be increased by increasing the thickness t (see Figure 6.4), but at the expense of reducing r. As t is increased, F increases but a decreases. The product Fa varies in some such way as shown in the graph in Figure 6.4.

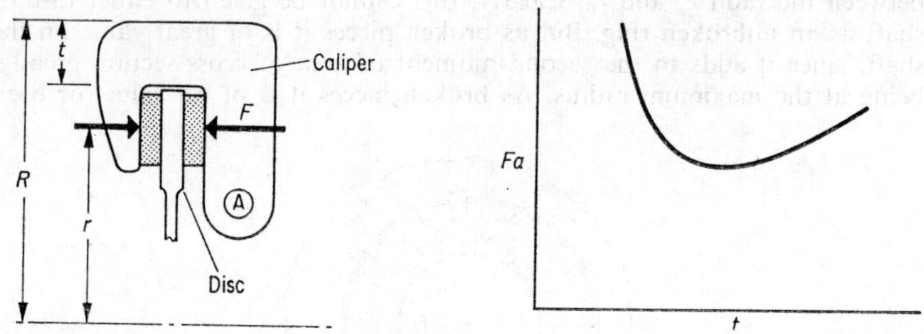

Figure 6.4 Disc-brake caliper disposition problem

It is possible to find an algebraic expression for Fa, quite simply, and to find the minimum by putting $\partial(Fa)/\partial t = 0$. It is better and easier, however, to plot Fa against t as in Figure 6.4; it gives the information we want, which is not the exact point at which the algebraic expression is a minimum, but the

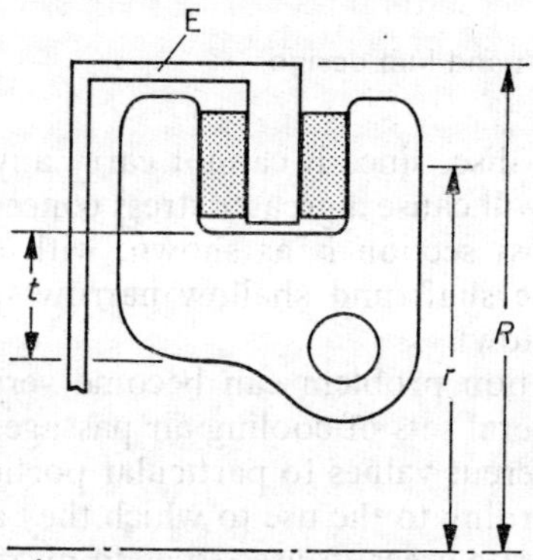

Figure 6.5 'Inverted' form of disc brake

way Fa varies with t to a practical level of accuracy. This can be used in a further optimization, say, taking into account material costs.

Notice that the commodity R to be disposed of is most valuable in the disc role when it is nearest the outside, whereas its value as caliper thickness t is roughly equal everywhere (it is slightly more valuable further in, due to greater curvature of the median surface). The logic of the disposition problem is clear; we ought, other things being equal, to put the disc outside the caliper as in Figure 6.5. Other things are far from equal, however, so we have to return to the original scheme.

We can also regard this excursion, which proves fruitless but might not have been, as an example of breaking the link in a logical chain. The link is that a can be reduced by increasing t only at the expense of increasing F. The weakness in the link is that it is only true if the disc is inside the caliper. It is a very useful practice to build up logical chains fairly rapidly, without, as it were, thoroughly testing each link, and then subsequently to re-examine them (section 9.5). An example occurred earlier when the order SCIH was eliminated in the L.N.G. tanker study, and later reinstated (section 2.5).

Finally, we could regard this as a case of inversion (section 9.4). All these three kinds of thinking:

(a) disposition and allocating bits of space to the uses for which they have the most value,
(b) logical chain making and breaking, and
(c) inversion, as part of a systematic variation of the configuration,

lead to the same idea, but the first is the best and surest. Notice that in both (a) and (b) the idea appears as an improvement, at least in one respect, while in (c) it comes only as one of a large number of random variations and without any *a priori* knowledge of the advantage it possesses. Both (a) and (b) are synthesizing steps, but (c) is merely a combinative routine, a mutation.

Notice that the arrangement of Figure 6.5 does not reduce the available space for the hub and wheel-carrier. It is true that a little of R is now taken for a new use, that of accommodating the cylindrical extension E of the disc. On the other hand, F is smaller, the radial extent of the pads is smaller, the arm of the bending moment which F exerts on the thickness t is smaller, and these effects should reduce the radial extent of the entire caliper more than enough to offset the cylindrical extension. Indeed, when the position of the pin A is considered, the space restrictions in the centre should be eased (section 7.2).

The new arrangement has at least three disadvantages:

(1) the removal of dirt and large flows of water from the 'disc-drum' presents problems,
(2) the 'disc-drum' is expensive, and
(3) the axial extent is increased: these are probably enough to eliminate it.

6.5 Alternator rotor

Consider the section of an alternator rotor shown in Figure 6.6. It is simply a rotating electromagnet with two poles, energized by a coil wound round the magnetic axis in slots cut in the rotor. The centrifugal loads on the rotor tend to burst it, and this limits the diameter, since the speed is fixed at 3000 rev/min by the need to generate at 50 Hz (cycles per second). Different designs of rotor will have different masses and strengths, and so slightly different limiting sizes, but for the moment the diameter will be taken as fixed.

Cross-sectional area is the commodity we have to dispose of to best effect, and it must be shared between the following functions:

s strength, to resist bursting,
f flux path, for the magnetic field,
c conductors, for the exciting current,
i insulation, and
k cooling (by water or hydrogen, say).

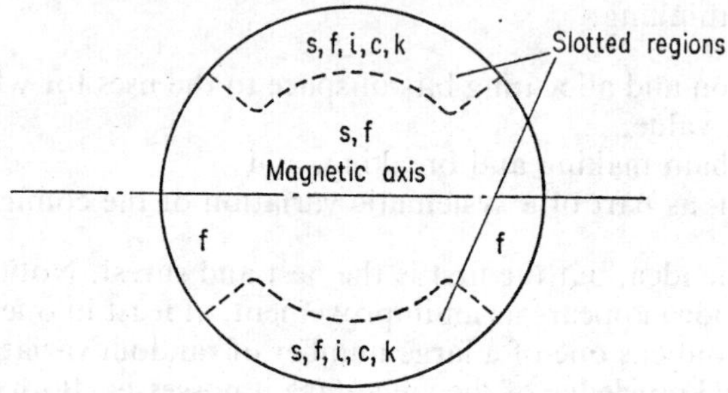

Figure 6.6 Alternator rotor (1)

Not all these functions are required in all areas equally, and in Figure 6.6 the regions in which various groups are most important have been shown by broken lines—broken because the boundaries are indistinct. All five functions are competing for space in the slotted regions, a section of which is shown in Figure 6.7. It has the conventional parallel slots, giving a 'tooth' which is

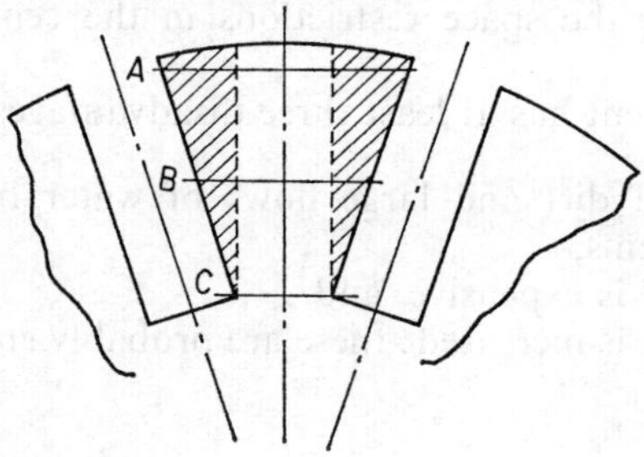

Figure 6.7 Alternator rotor (2)

wider at the tip than the root. However, there is a flux 'bottle-neck' at the narrowest part of the tooth C, and so the triangular areas shown shaded are wasted as far as function *f* is concerned—the flux through the tooth is limited (practically speaking) to the saturation flux through C, and could almost equally well be carried by a parallel tooth of the same root width, as shown by the broken line.

As far as function *s* is concerned, the shaded areas are again wasted, for the centrifugal load is a maximum at C, and less at wider sections such as A and B. We conclude therefore that it is the *teeth* that should be parallel, not the slots. True, there is a problem of transferring the centrifugal loads on the conductors into the teeth, but this merely complicates the argument without invalidating the conclusion.

A.E.I. made a change in this direction in 1968 in their 660 MW design. Instead of using a parallel slot as in their previous alternators, they introduced a slot tapered at the bottom but parallel at the top, giving a gain in conductor area of 16 per cent. While this is less than a slot tapered all the way would give, it leaves a little extra metal on the sides of the tooth tip in which to cut the wedge grooves which retain the conductors.

As in all such cases, there arises the question of how to distribute the fresh supply of cake—A.E.I. increased the ampere turns by only 8 per cent, using the other 8 per cent to reduce the current density, keeping the rotor copper losses unchanged though the rating of the alternator per metre of length was greater.

6.6 L.N.G. tanker

In this section a problem of disposition in three dimensions is studied, the design of the cargo section of a liquid natural gas tanker. One of the simplifications made is that the hull is assumed to be rectangular over the length containing the cargo tanks, although for reasons explained in section 3.3 the lines of these ships are rather fine.

The tanks will be assumed to be of the free-standing type, that is, (CS)IH or (CS)IH-SCIH hybrids (see section 2.5), so that for the purposes of this study they are large solid objects that contract on filling with the liquefied gas (at about −160°C). Their shape and size we are free to choose, but it must be remembered that they may have to stand internal pressures of up to about 0.1 N mm^{-2}, and that each tank will require a certain amount of equipment, valves, pumps, instruments, and so forth, so that a large number of small tanks would be expensive for this reason alone.

The available commodity is the space S inside our long rectangular box, and it is to be put to four uses—accommodating structure, insulation, L.N.G. tanks and other tanks (Figure 6.8(a)). The other tanks are to contain fuel oil and seawater ballast, and as they have no special requirements it seems sensible to leave them till last and fit them into any space that is not better used for another purpose.

Now L.N.G. is very light (specific gravity = 0.42), so that there is no possibility of overloading and we should try to devote as much of S to tank space as possible. The insulation must cover all the superficial area A of the tanks, to a thickness, t, say, (satisfy yourself that t should be uniform), so that an amount At of S is devoted to this purpose. If we have a rough idea of A, the quantity of L.N.G., the speed of the ship, its cost, and the journey length, we can estimate roughly the economical thickness t_e, bearing in mind that

(a) insulation costs money,
(b) space occupied by insulation cannot be filled with L.N.G.,
(c) natural gas costs money to buy and to liquefy,
(d) the amount of L.N.G. evaporated on the journey is roughly inversely proportional to t,
(e) the boiled-off gas can be burnt in the boilers.

This estimated thickness t_e from the rough calculation will be a useful figure in our subsequent studies; 200 mm is a typical figure.

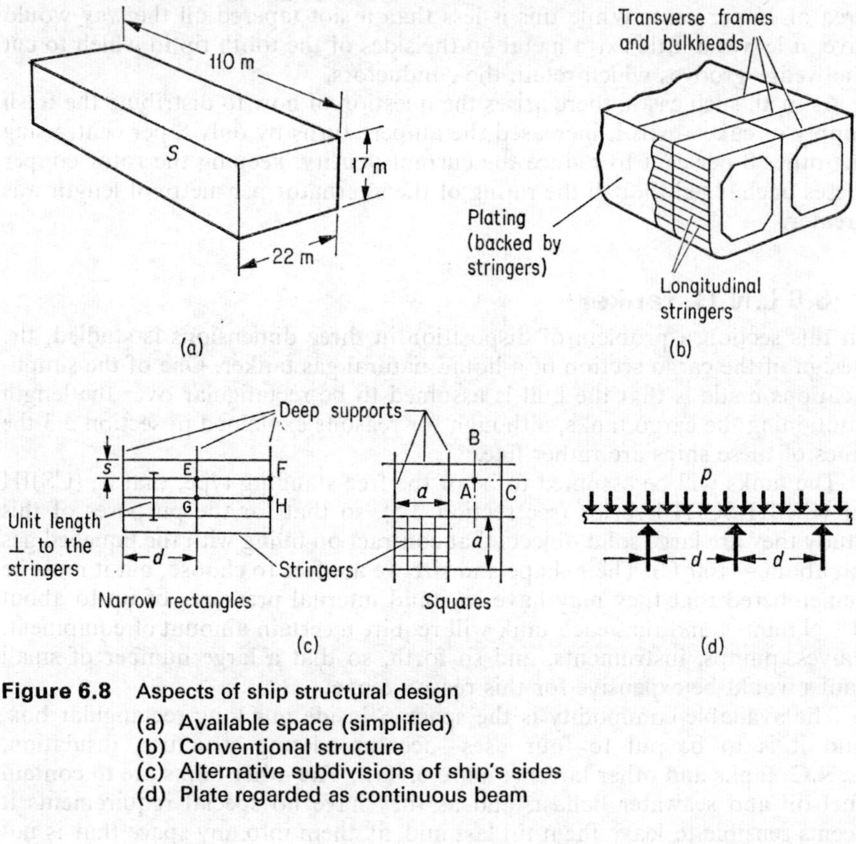

Figure 6.8 Aspects of ship structural design
(a) Available space (simplified)
(b) Conventional structure
(c) Alternative subdivisions of ship's sides
(d) Plate regarded as continuous beam

Suppose the effect of a change in design were to increase A by 10 per cent. To avoid increasing the rate of boil-off, the thickness t would have to be increased by 10 per cent also, and hence the amount of insulation by about 21 per cent. Both the capital costs and the cost of loss of capacity incurred due to insulation would go up by 21 per cent (of course, we should really reoptimize, accepting a slightly higher boil-off, slightly increasing ship speed and so forth, and then the increase in total cost would be slightly less than 21 per cent of the *insulation* costs; see section 3.3). It is clear, however, that there is a strong incentive to keep A as small as possible, which is practically the same as making the ratio of A to the volume V of L.N.G. a minimum.

This does not imply spherical tanks. It is true that a sphere has the lowest ratio of A to V for a given value of V, but it has the same ratio of A to V as the circumscribed cube, which has, of course, a greater value of V. If we look at our box (Figure 6.8) the biggest sphere we can put in it has an A/V ratio of $0.35\ m^{-1}$, but the box itself has an A/V ratio of only $0.23\ m^{-1}$. To keep down A it is on the whole more important to have big tanks than to make them the best shape. Moreover it can be seen that spherical tanks will not pack well into the available space. Rectangular tanks are not the best for resisting pressure, however.

So far we expect our design to have a small number of large holds lined with insulation and containing free standing tanks. It is reasonable to expect an A/V ratio of $0.3\ m^{-1}$, which with $t_e = 200$ mm means that about 6 per cent of the hold space will be occupied by insulation.

Having developed a little insight into the tank and insulation aspects, we turn to the structural ones. There are two principal strength requirements: to resist bending moments in the ship (of which we saw a little in section 4.2) and to resist the inward pressure of the sea. A few quick sums will soon show that the first is never likely to be difficult in this kind of ship, and so only the inward pressure is going to influence our design much.

The fraction of S which is likely to be occupied by steel structure is only about 0.01 (a quick calculation would show this), so that it is not this volume that matters, it is the enclosed or embraced spaces in the structure that cannot be used for cargo tanks (this is a link in the argument which will be reexamined later). We must look for schemes which minimize these lost spaces. When they are found, we must optimize the *scantlings* (metal thicknesses, stringer sections and so on; scantlings is a useful concept) for the lost space can be reduced by increasing the scantlings.

6.7 A digression: ship structures

Before searching for a minimum lost-space design, let us see how the ordinary ship resists sea pressures on its large flat sides (the bottom is a different matter). The surface is divided into smaller areas by bulkheads and deep web frames: the smaller areas are bridged by plates backed by stiffeners (Figure 6.8(b)). Let us now examine the logic of the shape of these smaller areas.

Figure 6.8(c) shows two extreme patterns of smaller areas—long thin rectangles of breadth d and squares of side a. Let us assume in comparing these two arrangements that there is no cost involved in running either the deep supports or the stringers through one another, and that the costs of both will be simply proportional to their lengths. Further, let us assume that the stringers have the same section and pitch and the plate the same thickness in each scheme, so that the section modulus z per unit length perpendicular to a set of stringers is the same in both.

Now in the thin-rectangle scheme we can regard the stiffened plate as a wide continuous beam over supports of pitch d, as in Figure 6.8(d), subject to a pressure difference p. In such a beam the maximum bending moment per unit width comes over the supports, and is $pd^2/12$. If f is the allowable stress, we choose d so that

$$\frac{pd^2}{12} = zf,\ d = \sqrt{\left(\frac{12\,zf}{p}\right)} \tag{6.1}$$

The next step is to carry out a similar rough calculation for the scheme with squares, which we shall do by making the bold assumption that at each point such as A in Figure 6.8(c) the pressure load is carried half by the stiffened plate acting as a beam along AC and half by it acting as a beam along AB. Then along AC, say, the plate/stiffener complex is a continuous beam of modulus z per unit width, supported at a pitch a and subject to a pressure difference of only $p/2$, so that

$$a = \sqrt{\left(\frac{12zf}{\frac{1}{2}p}\right)} = \sqrt{(2)}d$$

We can now compare the schemes. The length of deep support per unit area is $1/d$ in the rectangle scheme, and $2a/a^2 = 2/a = \sqrt{(2)}/d$ in the square scheme, so the latter requires 40 per cent more deep support and 100 per cent more stringer. True, we need to carry out a proper optimization to make an exact comparison, but this simple reasoning shows such a large advantage to the rectangle scheme that we run no risk in discarding the square. Two considerations reinforce this view: firstly, the cost of the intersections will be high and they must weaken the structure; secondly, the crude treatment of plate bending in the square overestimates its strength, as may be seen by the following argument.

Consider the two 'beams' AB, AC in Figure 6.6(c), where beam AB runs through the centre of a square and AC is near a support. If the two were independent and each carried just half the load, the deflection of AC at A would be much greater than that of AB. The actual deflections must be the same, so that AB must carry more load and AC less. In consequence, 'beams' such as AB carry $p/2$ at the centre and more than $p/2$ elsewhere, and the maximum bending moment must be higher than was assumed in fixing a.

By this short and simple study we have established a design philosophy for this particular problem that may be summed up as follows: to support a flat area against pressure, we use small beams consisting of a plate with stiffeners running one way, laid across deep supports (or larger beams) running the other way, rather as wooden planks are laid across joists in the floor of a house. Notice that the stiffeners themselves in Figure 6.8(c) define smaller rectangles EFGH over which the plate must support itself: clearly the plate here will work almost entirely as beams in the EG direction, and so the pitch of the stringers is determined by the plate thickness. If h is the plate thickness and s the stringer pitch,

$$f = \frac{\frac{1}{12}\,ps^2}{z\ (\text{plate alone})} = \frac{\frac{1}{12}\,ps^2}{h^2/6},\ \frac{s}{h} = \sqrt{\left(\frac{2f}{p}\right)} \tag{6.2}$$

Thus we have here a three-tiered structure, against the two-tiered structure of a wooden floor. A guide to the number of tiers which is appropriate in a given case is the *dimensionless structure loading coefficient*, another useful concept, which is p/f for this problem. A low p/f indicates a large number of tiers (three is large in this context) since it means a lightly loaded structure of relatively strong material. In such structures a little material is spread out through a large volume, and a complicated form is necessary to gather all the loads into the small solidity (section 7.5).

The rectangle scheme is easier to analyse than the inferior square one, and the load is carried in one way only as against two—this sort of directness and simplicity is a common feature of good solutions to structural design problems. It exemplifies a principle which might be expressed, 'better a stout belt or a solid pair of braces than a combination of belt and braces'.

6.8 L.N.G. tanker (resumed)

The next step is to see how all this relates to our special problem, which seems at present to be to provide the largest clear spaces we can inside the ship. Anything which may be regarded as a deep support will tend to encroach on such spaces, so d must be very large compared with an ordinary ship. This means that the overall depth of the stiffened plate q, say, must be very large, and as all this depth is lost space there is clearly a premium on a system of stiffening which gives the smallest q for a given span, d.

The question which follows, how do you dispose metal within an overall thickness q to enable it to span the largest d? is trivial. We put it in two equal plates separated by q (of course, some shear webs will be needed as well). We should not leave it at this, but quantify the value of various structures in this respect. The appropriate figure of merit F_M is

$$\frac{z}{q} \cdot \frac{1}{\text{volume of metal per unit area}}$$

$$= \frac{z}{q^2 \times \text{solidity}} \qquad (6.3)$$

which is another concept in line with choosing the best working point in the Tresca diagram (Figure 3.10) or efficiency of use (η_u: see section 4.5). If there

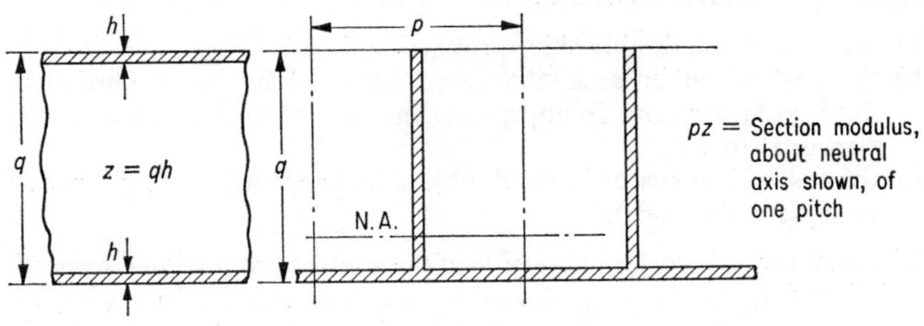

Figure 6.9 Section moduli per unit width (z) of structures

are two plates of thickness h, small compared with q (see Figure 6.9(a)) the solidity or ratio of metal volume to total volume is $2h/q$; z, the section modulus per unit length, is qh and the (dimensionless) figure of merit F_M is 0.5. By contrast the highest value for uniform stiffeners (Figure 6.9(b)) is about 0.18. Equation 6.1 shows d to be proportional to the square root of z, so that for the same q and the same weight of structure per unit area it should be possible to span 5/1.8 times the distance, or about 1.66 times the distance. Compared with the ordinary kinds of stiffening used in ship construction, there is a considerably greater improvement. But the ratio of about 1.66 is enough to be a strong inducement to adopt a double-skinned construction, especially as the holds will in any case require some sort of inner wall or lining to carry the insulation.

So far, the steps taken have pointed towards an arrangement of the sort shown in plan in Figure 6.10. Rectangular tanks of pitch d (where d is still our

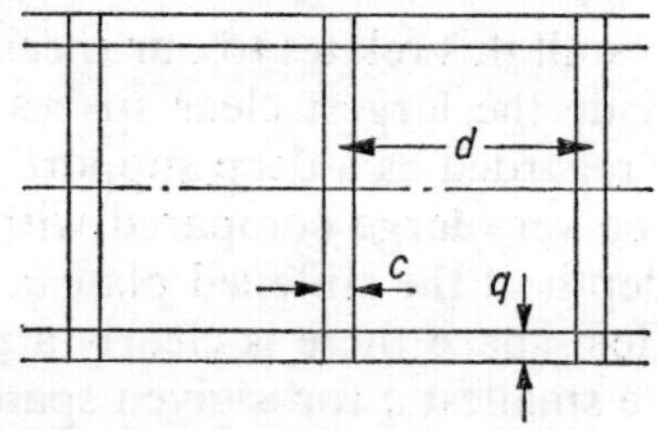

Figure 6.10 Optimization of rectangular tanks

rectangle width or span) are separated by double bulkheads ('cofferdams') whose thickness c is practically fixed (effectively it includes two layers of insulation and the distance between the two layers of plates, which is largely dictated by the access needed during building and after). The double skin has an overall thickness q and a section modulus per unit length qh, and the maximum bending moment in the skin is proportional to d^2, so that the allowable stress imposes the condition

$$qh = \text{constant} \times d^2 \tag{6.4}$$

The principal variable costs are:

(a) the cost of the double skin, proportional to h,
(b) the cost of cofferdams, the space occupied by the cofferdams, the fixed costs per tank (pumps, etc.) the cost of tank ends, etc. all proportional to $1/d$,
(c) the cost of lost spaces in the double skin, proportional to q (but subject to some deductions).

The total variable cost thus consists of three terms proportional respectively to h, $1/d$, and $1/q$, and by equation 6.4 $1/d$ is proportional to $1/\sqrt{(qh)}$. By a simple optimization on the lines of section 3.5, the optimum values of q, h, and d can be found.

Now is the stage to re-examine some of the links in the argument. Some of the questions that arise are:

(1) should the holds be clear space?
(2) need stringers and deep supports involve lost space?
(3) should the span *d* be fore-and-aft or circumferential?
(4) would the saving on some form of tank more suitable for resisting internal pressure outweigh the extra lost space (if any)?

To keep this study to a suitable length the treatment of these topics must be brief.

(1) The holds might be traversed by part of the structure, say, struts from side to side; such structures would have to be surrounded by insulation and accommodated in tubes passing through the tanks, increasing the boil-off rate and reducing the space available for cargo. It would not be possible to

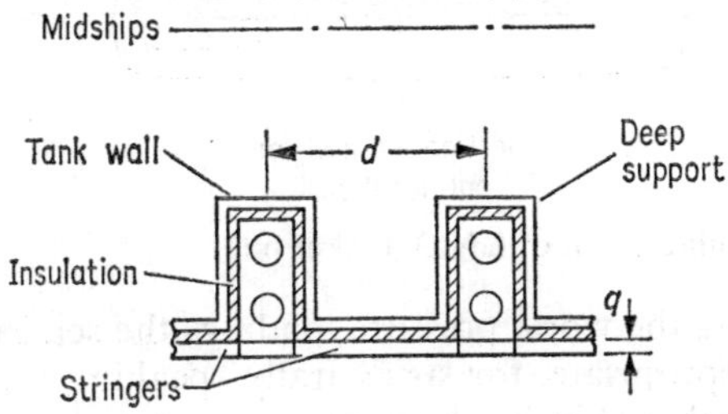

Figure 6.11 Tank following round deep supports

prefabricate the tanks on shore, and their construction would be more expensive. A differential study would suffice to eliminate such schemes, if there were any doubt.

(2) This is a similar but more practicable suggestion. The thickness of insulation aspect, apart from the complication, is enough to rule out the possibility of fitting the tank in between the stringers, but following round the deep supports, perhaps in the way shown in Figure 6.11, has some attractions. Increased tank and structure costs, however, tip the balance strongly against such schemes, even though the fraction of *S* devoted to cargo can be slightly increased.

(3) So far we have assumed deep supports running transversely. Deep supports running fore and aft must also be considered, otherwise the cardinal principle of 'never make an arbitrary decision' (section 9.2) would be flouted. If we have any intermediate longitudinal deep supports, they divide the smallest dimension of *S*, which is clearly the wrong thing to do when the object is the biggest possible tanks. Thus longitudinal deep supports are only viable if there are none, i.e. if the economic *d* from the earlier optimization approaches 17 or 20 metres and we can without excessive structural costs do without deep supports apart from those the bottom and deck provide for the sides and those the sides provide for the bottom and deck.

Question (3) leads to the idea of a double-skinned construction without any deep supports as such, i.e. what is effectively a double-walled tube of rectangular cross section, as in Figure 6.12. A few cofferdams are still needed, for safety when partly flooded etc., but these can be so far apart as to afford no useful support over most of the length. A thin slice of ship, such as that in Figure 6.12, if well away from the nearest cofferdam-bulkhead, must be

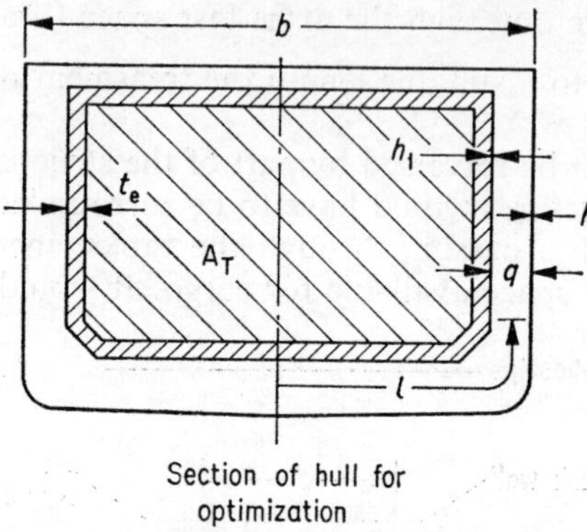

Figure 6.12 2-D optimization of L.N.G. tanker hull

capable of sustaining the worst pressure loadings the sea can exert on it. The idea of a slice is appropriate, for structurally speaking we can imagine ourselves cutting off such a ship by the yard.

Having arrived at this concept, the first thing to do is a quick sum to find suitable values for q and h. When this leads to reasonable results, a new problem of disposition is posed, which may be expressed as follows:

Given the section shown, to choose q, h_1 and h_0, which are functions of l, the peripheral distance from the keel, such that:

$$\text{shaded area } (A_T) - \text{structure cost per unit length} \tag{6.5}$$

is a maximum, subject to acceptable stresses, where of course structure cost is to be made commensurate with cargo cross-sectional area (see section 3.2).

The problem as stated is of the brachistochrone type (see section 3.10), but in practice it is only acceptable to vary q stepwise at a corner and h_1 and h_0

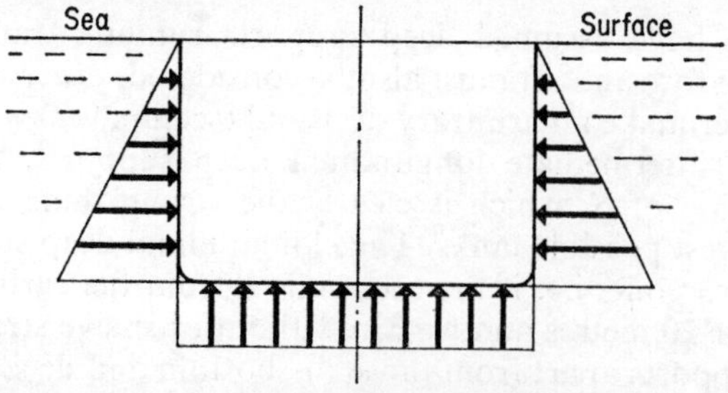

Figure 6.13 Loading condition in way of wave crest

stepwise at suitable intervals (i.e. at junctions between plates). The critical loading condition will be something like that shown in Figure 6.13, where the section concerned is awash (the pressure on the bottom is shown decreased by the downward load of the tank and contents). This problem can be solved exactly enough for practical purposes with quite a modest expenditure of effort, especially since it is possible to use an A.L.O.P. (section 3.13) to relate q and h.

One further complication has not been mentioned so far, and that is the question of stability. It is clearly advantageous to increase the depth of the ship as much as possible, and a limit is set by the need to maintain an adequate metacentric height. Free surface effects of the liquid natural gas enter into

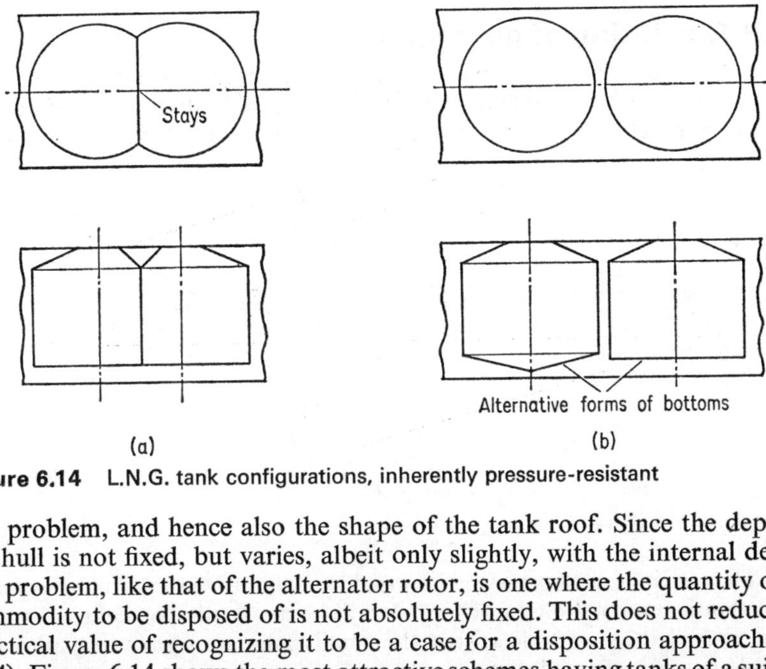

Figure 6.14 L.N.G. tank configurations, inherently pressure-resistant

this problem, and hence also the shape of the tank roof. Since the depth of the hull is not fixed, but varies, albeit only slightly, with the internal design, this problem, like that of the alternator rotor, is one where the quantity of the commodity to be disposed of is not absolutely fixed. This does not reduce the practical value of recognizing it to be a case for a disposition approach.

(4) Figure 6.14 shows the most attractive schemes having tanks of a suitable form to resist internal pressure. The 'volumetric efficiency' is rather low, especially in Figure 6.14(b), but the tank cost is also low. In the 'double bubble' form of Figure 6.14(a) the best volumetric efficiency, about 8 per cent higher than for simple cylinders, is had when the length is about 1.5 diameters, which is probably the greatest length that would be acceptable from the safety standpoint. With cylindrical tanks the structural behaviour of the hull falls into the 'belt and braces' category of the scheme with square areas of stiffened plate.

The French *Jules Verne* has cylindrical tanks of 9 per cent nickel steel, while the British *Methane Princess* and *Progress* have rectangular tanks of aluminium alloy.

This very compressed study of the disposition problem posed by the design of L.N.G. tankers with self-supporting (free-standing) tanks will be taken no further. Enough has been done to show the general lines of development. A noteworthy feature is that quantities like t/AK (where K is the thermal conductivity of the insulation), which may be regarded as the thermal resistance to earth, and the ratio of depth of hull to beam, need only be recalculated at intervals, rather as in the Newton-Raphson iterative solution of equations it is usually uneconomical to recalculate the slope dy/dx at each round. Another important point is the way in which at one stage a change occurs from a relatively simple three-dimensional problem to a tough two-dimensional one.

6.9 The design of gear teeth

In what follows it will be assumed the reader has some knowledge of involute gearing. The specific problem that will be treated is the design of ground involute spur teeth for a pair of gears of given diametral pitch and numbers of

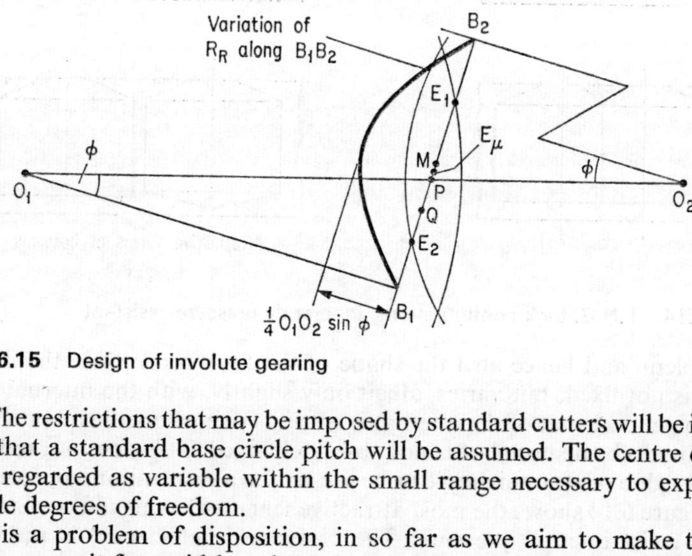

Figure 6.15 Design of involute gearing

teeth. The restrictions that may be imposed by standard cutters will be ignored, except that a standard base circle pitch will be assumed. The centre distance will be regarded as variable within the small range necessary to exploit the available degrees of freedom.

This is a problem of disposition, in so far as we aim to make the load capacity per unit face width as large as possible. It differs from the previous examples, and illustrates the wider class of problems that have only some of the characteristics of disposition.

Since the numbers of teeth and the base circle pitch are fixed, the base circle radii O_1B_1 and O_2B_2 are fixed (see Figure 6.15). By increasing or decreasing O_1O_2, the centre distance, the pressure angle ϕ may be varied. We can also choose the tip circle radii O_1E_1 and O_2E_2 to suit ourselves. Finally, we can choose the chordal thickness of one set of teeth at some suitable radius: the thickness of the other set will then be fixed by the need to have a suitable backlash.

We have thus four degrees of freedom of choice—ϕ, two tip circle radii and one tooth thickness. To simplify the problem, we shall suspend one of these by making the tip circles overlap by two modules (2/diametral pitch).

Suppose Q is a point of contact (Figure 6.15), then the radii of curvature of the flanks of the teeth at Q are B_1Q and B_2Q and the relative radius of curvature is

$$R_R = \frac{B_1Q \,.\, B_2Q}{B_1Q + B_2Q} = \frac{B_1Q \,.\, B_2Q}{B_1B_2} \tag{6.6}$$

where B_1, B_2 are the points of tangency to the base circles.

It was seen in section 4.8 that the load-carrying capacity of a line contact such as that between gear teeth might be expected to be proportional to R_R. The largest value of R_R occurs when contact is at M, the mid-point of B_1B_2, and is $\frac{1}{4}B_1B_2$ or $\frac{1}{4}O_1O_2 \sin\phi$. The variation of R_R along B_1B_2 is plotted in Figure 6.15; its form is parabolic, and symmetrical about M with zeros at B_1 and B_2.

The three principal sorts of failure we have to avoid, in the absence of shock loads, are:

(a) surface fatigue failure,
(b) 'bending' fatigue failures at the roots,
(c) scuffing failure.

To make the argument tidier, we shall ignore (c) for the moment, and see how the exercise of our three degrees of freedom of choice affects (a) and (b).

(a) Surface strength

Two of our three degrees of freedom of choice affect surface strength, ϕ, and whichever tip circle radius we regard as the independent variable, say, O_1E_1 (since $O_2E_2 = O_1O_2 + 2$ modules $- O_1E_1$).

For any point dividing B_1B_2 in a fixed ratio, R_R is proportional to $\sin\phi$, so that increasing ϕ increases the load capacity of a single tooth contact. As there will be a single pair of teeth in contact at and a little on either side of E_μ, the midpoint of E_1E_2, we might propose as a figure of merit for surface load capacity

$$(R_R)_{\min} \times \cos\phi \tag{6.7}$$

where the $\cos\phi$ factor allows for the higher tooth load produced by a more oblique pressure line and 'min' means minimum for a single pair of teeth in contact. To make $(R_R)_{\min}$ as large as possible we make E_μ fall on M, the midpoint of B_1B_2. Then $(R_R)_{\min}$ is virtually equal to R_R at M (perhaps 1.5 per cent less at most) so that the figure of merit becomes $\frac{1}{4}O_1O_2 \sin\phi \cos\phi$. We therefore make ϕ as large as possible.

The improvement to be had in the figure of merit compared with 'standard' 20° pressure angle gears by these two means, increasing ϕ and moving E_μ to, or near to, M, is of the order of 80 per cent in a typical case. The writer believes that the 'single pair' form of analysis given here underestimates the value of the higher contact ratio of lower pressure angles in the case of high-speed gears, where the assumption of constant wind-up in the teeth may be nearer the truth than that of constant tooth force. Nevertheless, the improvements in surface strength are at least of this order [28], perhaps because the effect of R_R is greater than Hertzian theory indicates.

(b) Bending strength

Figure 6.16 shows the effects of increasing the pressure angle on the bending strength of teeth. The effect of increased ϕ is

(1) to increase the tooth force as sec ϕ,
(2) to increase the radial component of tooth force,
(3) to decrease the bending arm h of the tooth force, about the midpoint of the base AB, and
(4) to increase the width of the base AB.

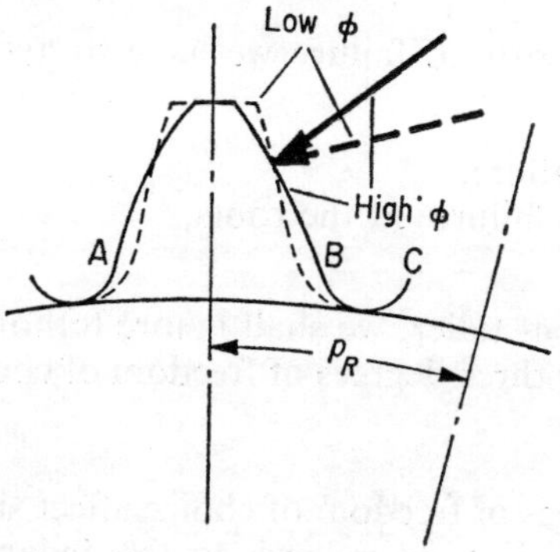

Figure 6.16 Effects of change of pressure angle

All but (1) are advantageous, and (1) is small and partly cancelled by (2). The dominant effects are (3) and (4), which greatly increase the bending strength. Again the indication is to increase ϕ, but here the necessary limit on the process is apparent (section 4.6). As ϕ is increased, both the lands at the tips of the teeth and the spaces between their roots become narrower, and both these effects must be watched. In particular, the root space presents a disposition problem in itself. Suppose p_R is the circular pitch of the teeth a little above the root radius. This has to be shared between the tooth base AB and the root space BC (Figure 6.16), and there will be an optimum ratio of division. If the root space is too narrow, the small fillet radius will more than offset the wider tooth base: the writer has used a ratio of 3:1.

In 'standard' gears, p_R is less for the pinion than for the wheel, since the circular pitches are equal at pitch circles and the centre lines of the teeth converge towards the centre more rapidly on the pinion. By shifting E_μ from the pitch point P towards M, the bending strengths of the wheel and pinion can be balanced. To some extent balance of bending strength can also be effected by varying tooth thickness, but clearly this cannot replace a sheer deficiency of p_R in the pinion. We can regard the whole bending-strength problem of the gears as one of disposing the combined p_R's between two tooth bases and two tooth spaces, except that as we move E_μ from P towards M the sum of the p_R's increases.

(c) Scuffing strength

It is even less possible to be accurately quantitative here than in the other cases, but the aims must be to keep down the surface stress and the sliding velocity. Since the sliding velocity at any point of contact Q is $(\omega_1 + \omega_2)$ PQ, where ω_1 and ω_2 are the angular velocities of the pinion and wheel, contact must be confined to a region as close to the pitch point as possible. The first of our two devices for improving surface strength, increasing ϕ, reduces the sliding velocities also, but the second, shifting E_μ to or towards M, increases the maximum sliding velocity.

We can summarize our findings with regard to the most important two of our three degrees of freedom as follows:

Kind of strength	Indications	
	Pressure angle ϕ	Position of E_μ
Surface	Increase as much as possible	At M, or as near M as possible
'Bending'	Increase to limit imposed by disposition of p_R or 'sharpening' of tips	On wheel side of pitch point P, sufficiently to balance bending strengths of pinion and wheel
Scuffing	Increase as much as possible	P

Non-Involute gears

It is possible to have tooth forms other than involute, with inherently higher surface strength. On a Hertzian basis it can be shown rigorously that the practical limit is almost reached by gears with circular arc profiles (Wildhaber, Novikov) [29].

6.10 Disposition of allowable stress

It often happens that the maximum stress in a component is the sum of stresses arising from several causes. In the suspension-bridge tower of section 3.4, the maximum stress is due to the combined effects of the downward load applied by the cables, the bending moment caused by the wind, lack of concentricity of the cable load and the axis of the tower, and possibly the fixing moments of the transverse members where a portal form of structure is used (see Question 10.11).

In such cases it may be appropriate to approach the design problem as one of disposition, where the allowable stress is to be shared out among the different terms in the most favourable way possible. The final criterion of the best design will be overall cost, but often some intermediate criterion, such as quantity of material used, can be substituted to make the problem more manageable.

In most of these problems, the severest stresses due to each cause will not occur simultaneously, or fatigue may be concerned together with a steady loading, and then simple addition of stresses will not do. Special techniques are then needed, such as a Goodman diagram or a statistical treatment, in order to decide what are permissible combinations of loadings.

The simplest case is when two stresses, arising from different causes and varying in different ways with the scantlings, add directly at some points. Consider a strap which is both bent and loaded in tension, and suppose the bending to be due to an imposed *distortion*, i.e. the strap is forced to deflect a given amount. Then the bending stress will vary directly as the thickness t. On the other hand, if the direct stress is due to an imposed *load*, it will be inversely proportional to t, and the combined stress will be of the form

$$At + \frac{B}{t} \tag{6.8}$$

where A and B are constants and the first term is due to bending and the second term to the direct load. Question 7.2 is of this sort, and an example is given on page 6 of reference 30.

6.11 Electromagnets

The design of electromagnets to produce intense magnetic fields involves some interesting disposition problems. Figure 6.17(a) shows the hollow cylindrical

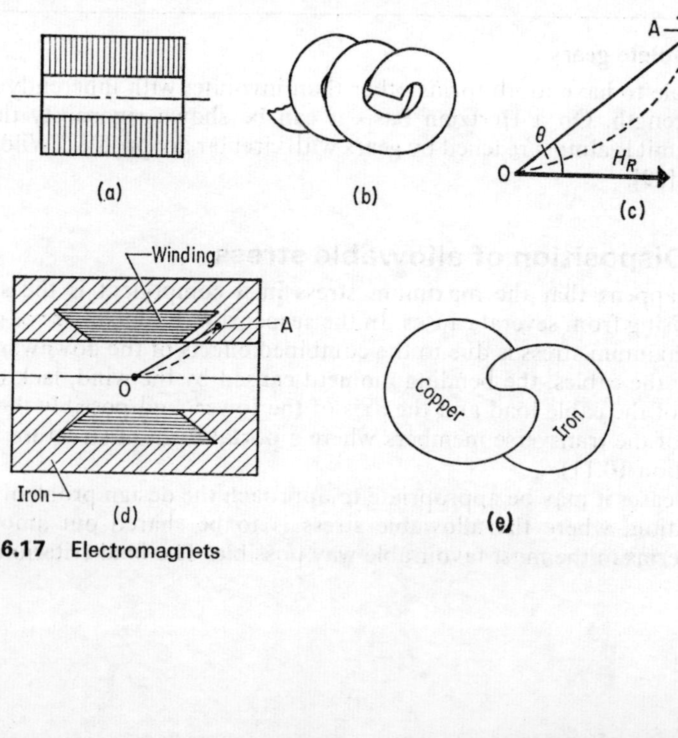

Figure 6.17 Electromagnets

coil of an air-cored solenoid. The portions of the coil nearest the axis produce more ampere turns per metre of conductor, so that an optimum distribution of current density is non-uniform (in fact, inversely proportional to the radius). This can be achieved by making the coil of a thin flat spiral conductor [31] as in Figure 6.17(b).

In an electromagnet using iron, the effect of the iron can be considered as that of minute magnets, or dipoles, lying along the magnetic field. In Figure 6.17(c), if O is the point at which we want to produce a strong field in the direction H_R, and A is any point in the iron, we can imagine a dipole at A producing a small increment of field at O, which will have a component along H_R. If we rotate the dipole in the plane of the figure, this component will vary, and it makes its maximum contribution in position U. In another position, V, it has no component, and when turned through 180° it opposes H_R. Thus we seek to arrange all the dipoles in the iron in the U position: difficult as this task may sound, the design by Bitter in Figure 6.17(d) achieves it. The working volume of the field is produced in the central hole in the iron, at O. The coil should have the same current density variation as before, i.e. inversely proportional to radius, but this time it is achieved by winding the coil in a single spiral from a tapered strip of foil, beginning in the centre with the narrow end. This yields the cylindrical coil with conical depressions in the ends shown in the figure.

Figure 6.17(e), which follows Laithwaite, is an idealization of an electromagnetic device—motor, magnet, choke—as a magnetic circuit linked with an electric circuit as closely as possible. Laithwaite [32] has an important concept—his 'goodness factor', which is a measure of the excellence of the linking of the two circuits. Goodness factor is an example of the sort of idea which springs only from design, and is so lacking in the engineering vocabulary (see section 4.5).

Questions

Q.6.1. Consider the design of a hotel bedroom to accommodate two single beds 0.9 m × 1.9 m, a wardrobe with drawers inside and double-hinged doors, 1.2 m × 0.6 m in plan, a vanitory unit 0.5 m × 1.2 m, two wicker armchairs, a table 0.6 m × 0.8 m and a luggage stand 0.7 m × 0.5 m. A window is to be provided in one short wall and a door in the opposite wall and the length of the room should be 1.5 times the breadth. There is a radiator below the window. Using sensible guesses where properly you should use ergonomic data, find the smallest size of room you would recommend, and show the arrangement you propose. (A room much deeper than 1.5 times its window wall length will seem gloomy because of poor penetration of daylight, unless the ceiling is excessively high, and a squarer room will give a shallower and hence less economical building.)

Q.6.2. Consider the disposition problem set by the design of cupboards for kitchens, especially by the range of heights of the articles stored, which should be assumed to be up to 320 mm, and the need for easy access to most items.

Q.6.3(2). A spur pinion has a circular pitch at the roots of the teeth p_R and 'semi-circular' fillets of radius r between the teeth. If the tooth chordal width at the root is b, so that $b + 2r = p_R$, find approximately the optimum value of r assuming that the bending stress in the tooth may be estimated by the usual beam theory, with a stress concentration factor K given by

$$K = 1 + 0.15\frac{b}{r}$$

Q.6.4(3). Figure 6.18 shows a sketch for a ball joint connecting two rods in a linkage. Regarding the design of such a joint as a problem of disposition of an

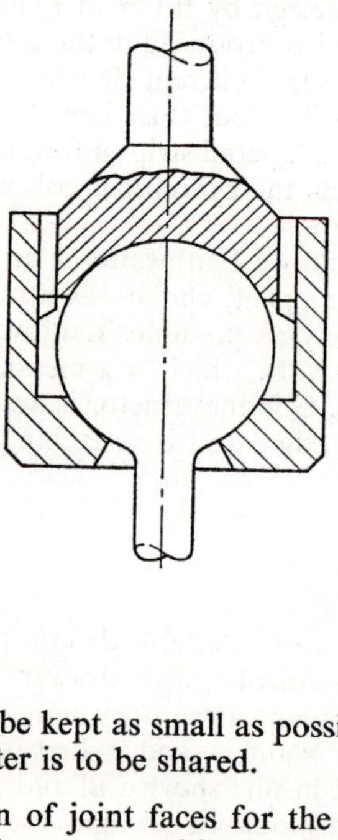

Figure 6.18 Ball-joint

overall diameter that is to be kept as small as possible, analyse the functions between which that diameter is to be shared.

Q.6.5. Consider the design of joint faces for the halves of connecting-rod big ends (see Figure 6.19 for an example) as a disposition problem. Make a functional analysis of the joint face, and examine the relative values of different regions of the available cross-sectional area for different uses.

Q.6.6. A heat engine working on the Rankine cycle and using a halogenated hydrocarbon as a working fluid extracts heat from a plentiful supply of hot water at T_1 and rejects it to the atmosphere at T_0. Analyse the functions to be performed by $T_1 - T_0$ and state the disadvantage incurred by reducing the fraction of $T_1 - T_0$ allotted to each.

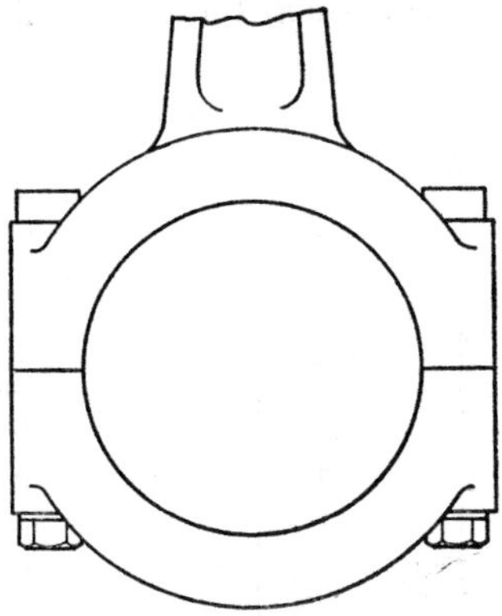

Figure 6.19 Connecting-rod big-end

Answers

A.6.1. About 3 m × 4.5 m. Some general principles are, that spaces for bedmaking, swing of wardrobe and room doors, etc., should so far as possible have more than one function, and that one large squarish clear space is much more convenient and comfortable than an equal area disposed in long narrow strips. It follows that the perimeter of the free space should be as small as possible relative to its area—in engineering jargon, the mean hydraulic diameter should be large. This short perimeter must therefore consist largely of bedsides, wardrobe and room doors, vanitory unit, etc.

A.6.2. Access to and removal of most items without disturbing the others is desired. Dividing the accommodation between shelves in the cupboard and receptacles on the back of the door, as in refrigerators, is a great help, but this is limited by the acceptable inertia and weight on the hinges. Shallow drawers suspended under part of the width of a shelf also help, as do intermediate shelves at some levels extending perhaps one third of the way from the back of the cupboard.

A.6.3. The factored 'bending' stress is roughly proportional to

$$\frac{1 + 0.15\, b/r}{b^2}$$

since the section modulus of the root section of the tooth is proportional to b^2. Substituting $r = \frac{1}{2}(p_R - b)$ (or using Lagrange) and minimizing (section 3.10) gives $b = 0.76\, p_R$ approximately.

This result is reasonable, but data are available for better treatments. Furthermore, complications of tolerances, preformed roots, etc., shift the optimum value of b downwards slightly.

A.6.4. The functions amongst which the overall diameter is distributed are (Figure 6.20(a))

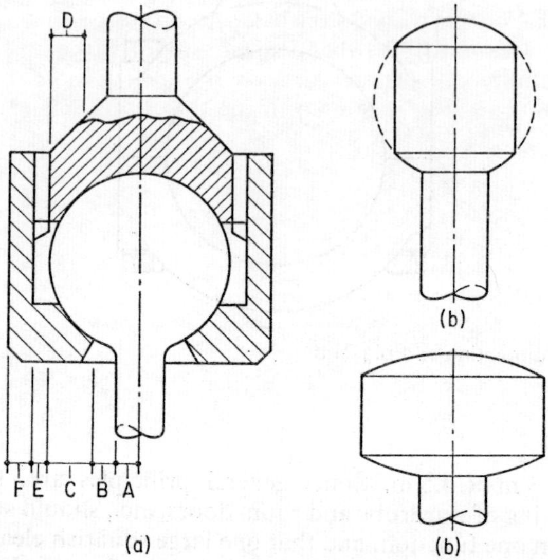

Figure 6.20

A ball rod
B freedom
C tensile bearing area
D socket end structure
E joint
F tensile structure

It is clear that D need not encroach on the diameter but the others must (except that E can be got round by extreme constructions such as a longitudinally split socket bolted through above and below the ball). It is also plain that it is better to use ball joints in compression where possible. In the front suspension of cars where the stub-axles are carried on vertical members (king pins) with ball joints top and bottom, these members are sometimes joggled at the lower end so that all the joints have the stem of the ball downwards. In this case the reduced frictional resistance to steering in the load-carrying bottom joints is important.

In a proper study, it would be desirable to add functions G (tolerances) and H (fillets and chamfers). If the freedom required is small enough, it may be worth using a slightly larger ball turned down round the equator (Figure 6.20(b)) or even the form (c) if a unique centre of hinging is not necessary.

A.6.5. The functions of the joint are:

(a) transmitting direct load
(b) transmitting shear load
(c) transmitting bending moment
(d) ensuring accurate positioning of one part relative to the other, i.e. providing a register.

Of these, (c) is the most exacting and (b) and (d) are intimately connected. The fulfilment of (a) and (c) by a joint of this sort depends on loading the faces together by a prestressed bolt or bolts, the function of which is that of a very stiff spring. For this reason, it is better in disposing the various functions in the available area to forget (a) and substitute (e), providing prestress. Now study the suitability of various parts of the available cross-sectional area (Figure 6.21(a)). The area is limited in the directions A and B by clearance

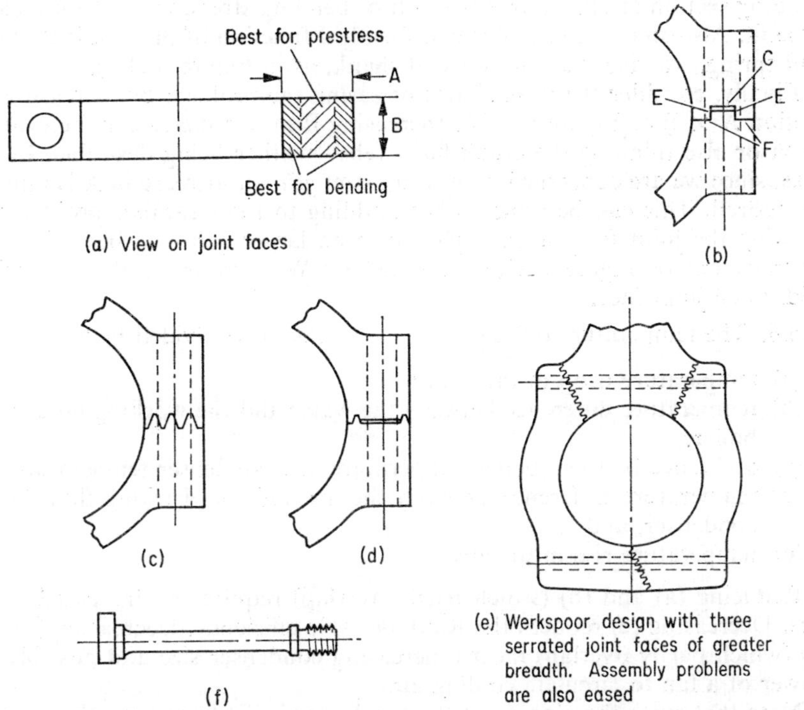

Figure 6.21 Joint faces of connecting rods

problems, but it will make sense to use the whole rectangle. Since bending occurs in both directions in the plane of the connecting rod, the prestressing material is best put in the middle and the bending function given to the 'flange' region, as in Figure 6.21(d). The bolt(s) will thus be put through the middle of the area. Coming to the shear-resistant and register functions, any part of the area will do. The stepped face of Figure 6.21(b) is worth consideration, provided that there is a clearance at C to ensure contact on the areas F which are well suited to take the bending. The shear-resistant function is adequately filled, but the maintenance of a register demands making the faces E and F all at once. A lighter, more solid job is given by serrations as in Figure 6.21(c). Finally, consideration might be given to removing the central serrations as in Figure 6.21(d), since the shear and register functions are adequately supplied by those remaining and the section stiffness in bending is scarcely altered. On the other hand, the same bolt at the same stress will give a higher prestress in the vital area.

Note that since the radius of gyration of the cross-section of the bolt is bound to be less than that of the joint face, it is best to free the bolt of bending and increase the preload—we aim at a strict separation of function, the bolt providing a direct prestress only and not contributing in its own right to the bending section at all. To free the bolt of bending stress, we make it a clearance fit. Also since we have decided the ideal function of the bolt is that of a stiff spring, we give it a long waisted shank, as in Figure 6.21(f).

Finally, consider the possibilities of easing the problem by increasing the section A × B of Figure 6.21(a). Increasing B means increasing the cylinder pitch or else thinning the crankshaft webs—neither being desirable. In any case, since we are concerned with beam strength, an increase in A is more to be desired. This can be done without adding to the clearance problems by skewing the joint face to an angle. An even larger gain can be had by the arrangement of Figure 6.21(e), adopted by Werkspoor, at the cost of an additional joint face.

A.6.6. The temperature difference $T_1 - T_0$ has to be divided between

(a) temperature drop in the water,
(b) temperature difference between the water and the working fluid in the boiler,
(c) difference between boiler temperature and condenser temperature,
(d) temperature difference between the air and the working fluid in the condenser, and
(e) temperature rise in the air.

Reducing (a) and (b) (which partly overlap) requires an increased boiler size. Decreasing (c) reduces the Rankine cycle efficiency. Decreasing (d) and (e) (which partly overlap) means increasing condenser size and possibly the power of a fan to circulate cooling air.

Note that since the air side of the condenser is likely to have the smallest heat transfer per unit area for a given temperature difference, we must expect (d) to be greater than (b) in an optimized design.

7
Kinematic and Elastic Design

7.1 Introduction

While it is not the purpose of this book to discuss special items in the engineering repertoire, the complementary ideas of kinematic and elastic design are so important as to warrant an exception.

The principle of kinematic design, which was expounded and used by Kelvin, may be stated as follows: 'When locating or guiding one body relative to another, use the minimum number of constraints'.

A simple and well-known example is that of the three-legged stool, which obeys the principle and always stands firm on its minimum number of three legs. A four-legged chair, however, has one more than the minimum number of constraints, and so will often stand on only three of its four legs. However, if the floor is flat and we adjust the length of the legs so that their ends lie very nearly in a plane, it will rock only slightly. If now we sit on the chair, the stresses our weight sets up will deform it so that all four legs are carrying some load. The four-legged chair is an elastic design, the load distribution in it depending on its flexibility and also on the perfection of its 'fit' to the floor, i.e. its freedom or otherwise from rock when under no load.

7.2 Spatial degrees of freedom: examples of kinematic design

For those who never were, or no longer are, familiar with the idea, some explanation of spatial degrees of freedom may be helpful.

An object has n spatial degrees of freedom if n co-ordinates are required to specify its position. Thus, a moving point in space has three degrees of freedom since it needs three co-ordinates, x, y, and z, say, to specify its position. A line AB of fixed length has five degrees of freedom; A and B would have three degrees of freedom each, corresponding to the co-ordinates x_A, y_A, z_A, x_B, y_B, z_B, but because the length AB is fixed,

$$(x_A - x_B)^2 + (y_A - y_B)^2 + (z_A - z_B)^2 = (\mathrm{AB})^2 = \text{constant}$$

and so if five co-ordinates are given the sixth is fixed. Another way of seeing this is as follows; if A is fixed, B moves on a sphere, where its position can be specified by only two co-ordinates, say, latitude and longitude. Hence the degrees of freedom are three for A, plus two for the latitude and longitude of B. The two points A and B have their degrees of freedom reduced from six to five when the distance between them is fixed. We can express this by saying that a link of fixed length imposes one simple constraint.

If a point is constrained to move on a line, where its position can be specified by a single co-ordinate such as the distance from one end, it has lost two degrees of freedom. Such a constraint is equal to two simple constraints.

A rigid triangle ABC has six degrees of freedom (three for each corner less three simple constraints, the constant lengths of the sides). A corollary is that any rigid body has six degrees of freedom, so that to support an engine in a helicopter, say, six simple constraints are the least which will suffice. Figure 7.1 shows such a scheme diagrammatically.

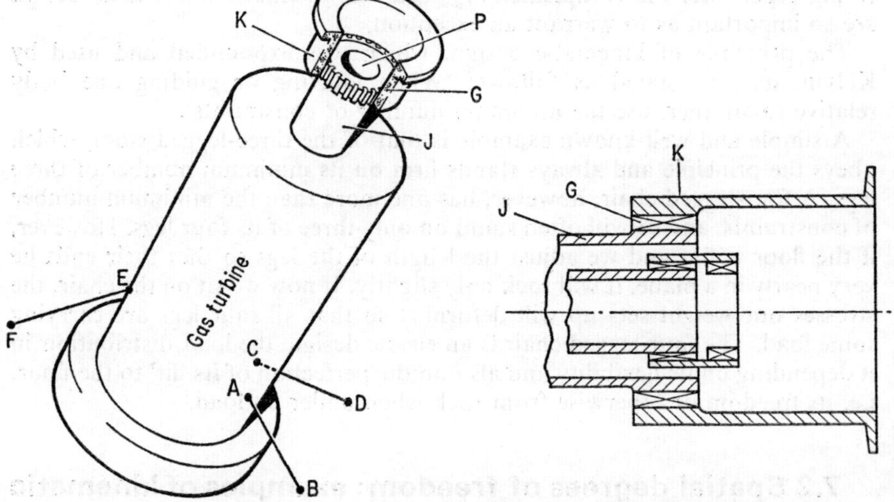

Figure 7.1 Kinematic mounting of helicopter engine

The engine is supported by three struts AB, CD and EF, freely pivoted at their ends, providing three simple constraints. At the upper end of the engine, there is a tubular extension J surrounding the drive shaft P. This extension slides in a tubular mounting K secured to a gearbox which is itself secured to the airframe. J is constrained to move on a line, the centreline of K, constituting two further simple constraints. Also J and K are provided with interlocking gear teeth, so that J is prevented from rotating: this provides the sixth constraint.

Notice that with this arrangement strains in the airframe do not manifest themselves in the engine, and vice versa. The loads in the struts and at K are statically determinate. An incidental advantage is that the path of the torque reaction, from the engine reduction gear casing to the toothed ring K on the aircraft, is the shortest practicable. Installation is simple—the engine is 'plugged' in at K and pins inserted at A, C, and E. The loads at A, C, and E are direct forces—no moments are exerted on the casing at these points.

The helicopter engine mounting is an example of kinematic design as just defined. There is a very similar kind of application that does not, however, always meet the condition of minimum constraints. It arises when a degree of freedom is deliberately added to a system to ensure that a desired balance of loads is obtained. For example, in the disc brake of Figure 6.4 it is highly desirable that the nipping force *F* between disc and pad should be the same on

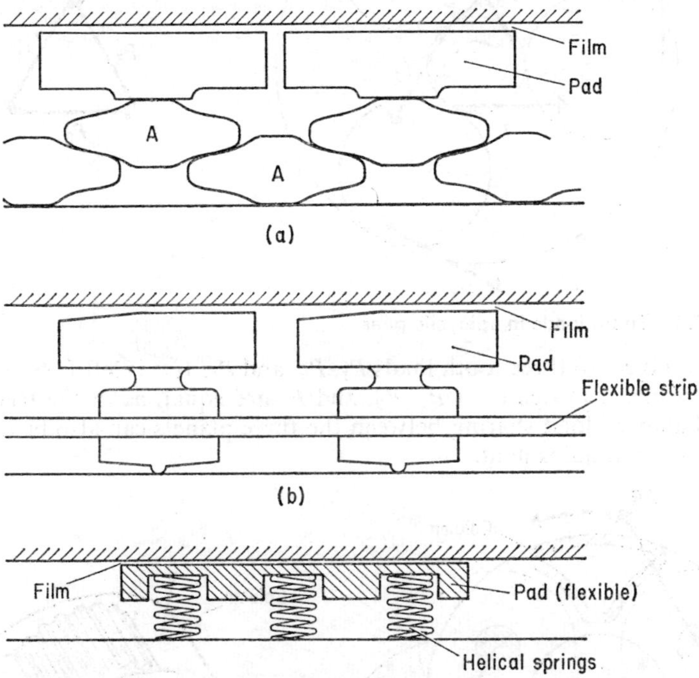

Figure 7.2 Backing of tilting pad bearings

each side of the disc. This can be arranged very simply by giving either the disc or the caliper a degree of freedom in a direction substantially parallel to the axis of the stub shaft. In Figure 6.4 the caliper is given a balancing degree of freedom by being pivoted about the pin A, which also forms the attachment to the wheel-carrier.

Further examples of load-balancing degrees of freedom occur in those surface plates which stand upon nested series of three-legged members such that all the points of support carry equal loads, and in the support system for some tilting pad thrust bearings (see Figure 7.2(a)). If friction and slight variations in geometry are ignored, the levers A will equalize the loads on all the pads.

An interesting case is that of epicyclic gearing. First consider a simple epicyclic gear, like that in Figure 7.3, as a plane static mechanism. The sun

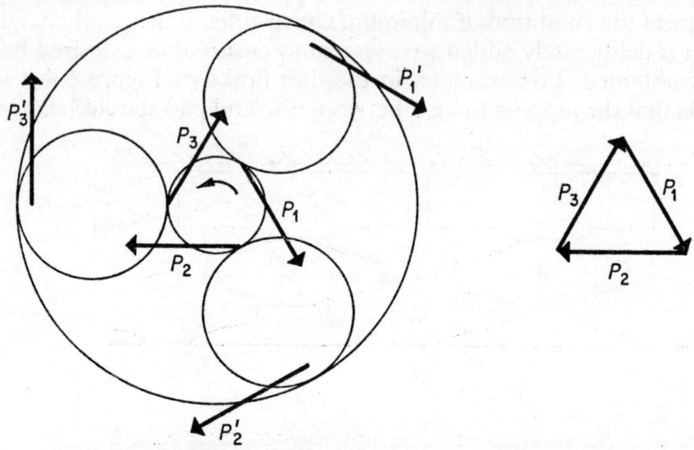

Figure 7.3 Tooth loads in epicyclic gear

pinion is subject to three tooth loads P_1, P_2, and P_3. If it is left free to float in its own plane, it moves until P_1, P_2, and P_3 are equal, as in the triangle of forces. Balanced load sharing between the three planets can also be achieved by letting the annulus float.

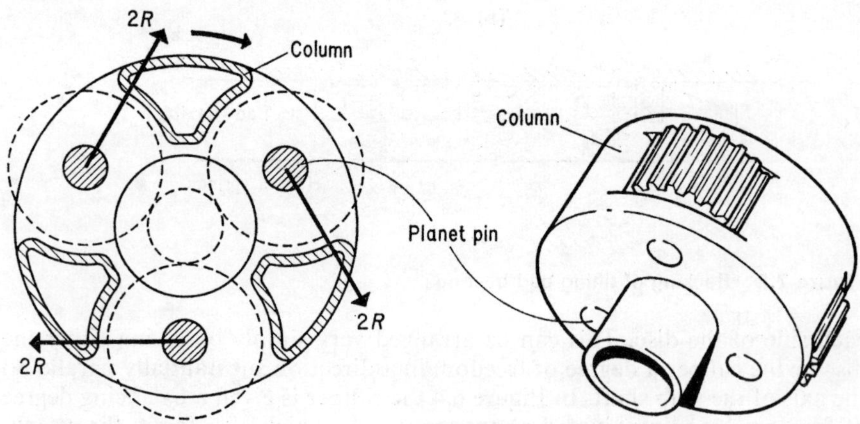

Figure 7.4 Twisting of spider of epicyclic gear

In the dynamic case, of course, the inertia of the sun, annulus and planets means that large alternating unbalanced loads can occur. Here the elasticity of the annulus gear may help, since the deflection under load can be of the order of 0.0015 of the radius.

Let us now consider the epicyclic gear as a three-dimensional device; imagine it being used as a reduction gear with the input to the sun pinion and the output from the spider. The spider has two end plates that support the ends of the three pins on which the planets run, and these three pins alternate with three columns that unite the end plates into a single structure (see Figure 7.4).

The tooth loads on each planet are reacted by two roughly equal forces R at the ends of its pin. The three forces R at the end remote from the output subject the spider to a torque $3aR$, where a is the pitch circle radius of the planet pins. With practical materials and construction the effect is to twist the spider, so that the pins lie in a slightly skewed position, and the planet gear teeth make an angle with those of the sun pinion and the annulus gear. The

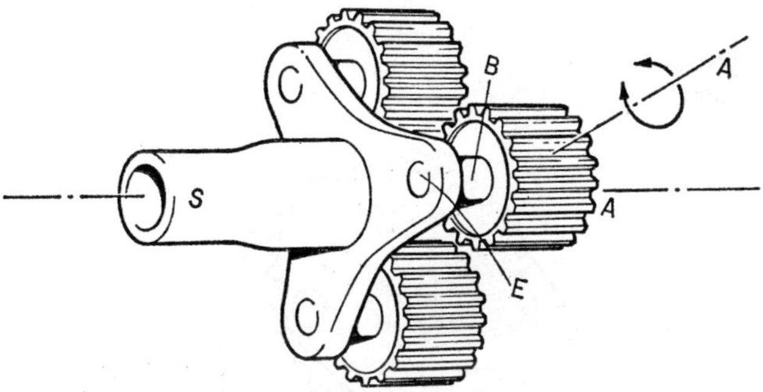

Figure 7.5 Kinematic design of spider and planets

practical results are heavy, expensive spiders and badly distributed loading on the teeth.

The solution to this problem is to introduce a further degree of freedom of the planet gears, so that they are free to align themselves with the sun and annulus. The arrangement of such a spider is shown diagrammatically in Figure 7.5. The output shaft S carries a single, relatively light, three-legged member E bearing three overhung pins B. The planets are pivoted to these pins in such a way that they are free to turn about an axis AA. (This probably implies they have another extra degree of freedom also, about an axis perpendicular to both AA and the axis of B.)

It is instructive to list the items from the engineering repertoire that would be suitable to provide this additional degree of freedom. They can be classified according to whether the degree of freedom is kinematic (i.e. based on relative sliding or rolling) or elastic.

Kinematic

(1) Self-aligning antifriction bearings, where the rolling elements provide for both the rotation of the planet gear and its alignment—spherical roller bearings or self-aligning ball bearings. The load capacity of the latter would normally be inadequate. (Why? see section 4.9.)

(2) Separate antifriction bearings providing a single degree of freedom. These could not normally go inside the pin, for then they would be too small to carry the load; the alternative is a gimbal or yoke arrangement, which is likely to prove clumsy.
(3) Spherical hydrodynamic (or hydrostatic) bearings providing for both planet rotation and planet alignment.
(4) Spherical (or cylindrical) hydrostatic bearings allowing for planet alignment only.
(Notice that stationary, non-hydrostatic plain bearings, allowing only for alignment, would give too much Coulomb friction.)

Elastic

(1) Pins,
(2) Straps,
(3) Diaphragms,
and some others.

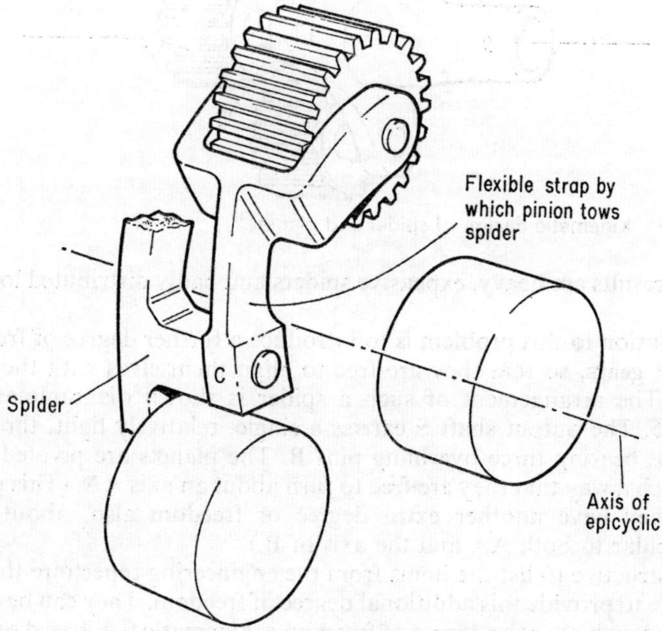

Figure 7.6 Strap mounting of epicyclic planet

Figure 7.6 shows a strap arrangement that could be converted into kinematic means (2) by a bearing at C. Figure 7.7 shows a diaphragm arrangement (from the Rolls-Royce Gazelle engine) and Figure 7.8 shows a pin arrangement.

Notice that in the arrangement of Figure 7.7 the drive is fed into the mid-length of the sun pinion by a quill shaft. This is done to minimize the effects of

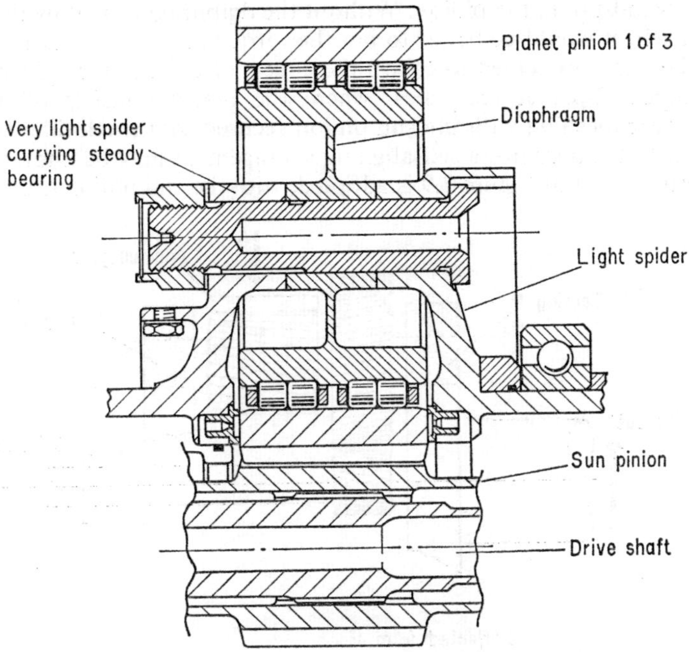

Figure 7.7 'Gazelle' reduction gear. (By kind permission of Rolls-Royce Ltd.)

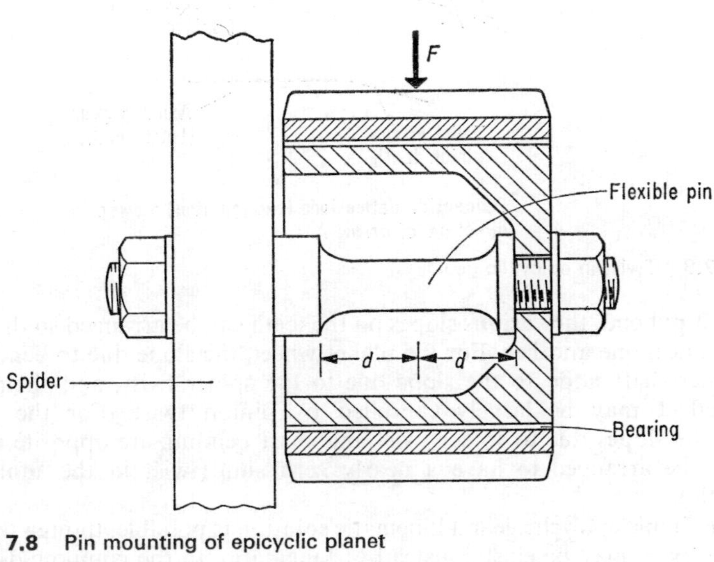

Figure 7.8 Pin mounting of epicyclic planet

torsional 'wind-up' in the pinion. Without the diaphragm to allow the planet to align itself, it would be better to put the torque into the end of the pinion so that the wind-up helped to cancel the effects of twist in the spider.

In compound-planet epicyclics, where the planet consists of a large-diameter gear meshing with the sun pinion secured to a small diameter gear meshing with the annulus, a self-aligning arrangement may be impracticable. If the arrangement of Figure 7.9 is adopted, with the sun pinion train nearest

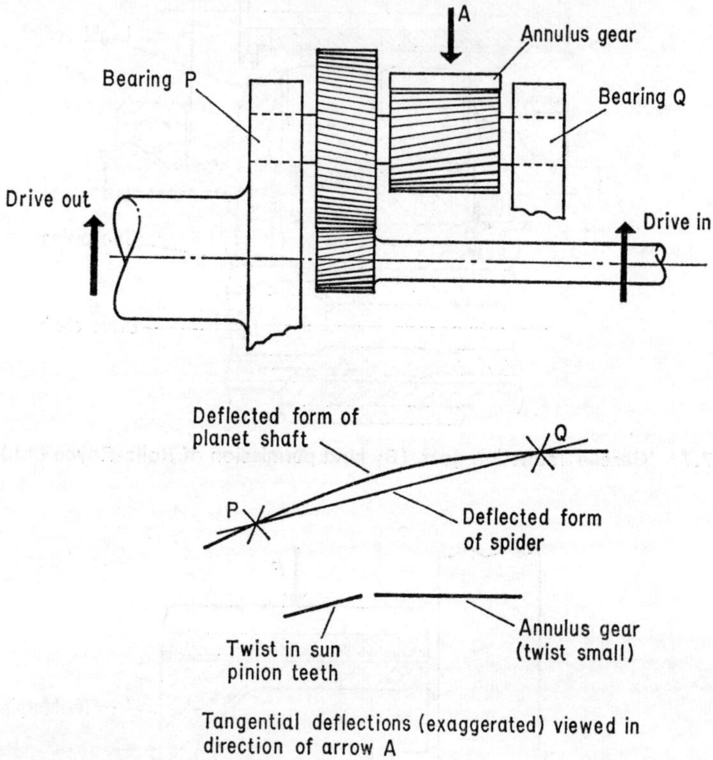

Figure 7.9 Twist in epicyclic gears

to the output end, the various slopes on the teeth can be arranged so that they nearly cancel one another. For the planet wheel, the slope due to bending in the planet shaft adds to the slope due to the spider twist, and their combined effect may be largely cancelled by pinion twist. For the planet pinion, the slopes due to spider twist and shaft bending are opposite in sign and can be arranged to have a nearly zero sum (twist in the annulus is negligible).

In the simple epicyclic gear a kinematic solution is possible, though the self-aligning 'pair' may be elastic instead of kinematic. In the compound-planet epicyclic gear the solution suggested is a characteristic piece of elastic design.

When neither of these kinds of solution is feasible (see next section), recourse must be had to making the teeth depart from the straight in the unloaded state, by modifying the helical angle or using longitudinal relief in addition to the usual small amount at the tooth ends. Modifying the tooth form in this way will work properly for only one load. Tooth matching is done this way in marine reduction gears where, owing to the long slender pinions, the problem is especially severe and the scope for more refined solutions is small.

7.3 Elastic design and elastic pairs

In the last section it was shown how the load distribution along the teeth of a simple epicyclic can be made nearly uniform by what is virtually a kinematic design using an elastic hinge instead of a kinematic one. The load distribution along the teeth of an epicyclic with compound planets can be made nearly uniform by suitable choice of the flexibilities of the components and a folded arrangement of the trains; this second problem might be solved kinematically also, by hinging each gear of the compound planet independently and connecting them by a flexible quill shaft, but such an arrangement would almost certainly be expensive, clumsy and heavy.

Now this use of elastic components in order to ensure fairly uniform tooth loads in the compound epicyclic planet is not an approximate realization of some kinematic design, but a fundamentally different idea of a kind we shall call 'elastic design'. It represents a diametrically opposite approach.

Kinematic design aims at a statically determinate system, where the forces in all the parts are fixed by conditions of equilibrium alone. Elastic design on the other hand aims at a redundant system, where the forces in the parts cannot be determined by considerations of equilibrium alone but depend upon their relative stiffnesses, i.e. equations of compatibility are involved. However, in elastic design the stiffnesses are regulated so that the compatibility conditions lead to desirable distributions of loads and forces. We might say that the principle of kinematic design is: 'never start a fight'. On this analogy, the principle of elastic design is: 'never start a fight unless you pick the contestants so that the outcome will be to your liking'.

One important difference between kinematic and elastic design is that small variations in dimensions do not affect load distribution in kinematic systems but they do in elastic ones; as a consequence the flexible elements of the latter must be able to absorb the effects of manufacturing tolerances without either unbalancing loads to an unacceptable extent or exceeding allowable stresses. As will be seen later, high flexibility means a large component, so that in elastic design it is sometimes necessary to choose between awkwardly large elements or rather stringent tolerances.

Designs such as the self-aligning planet of Figure 7.7, which may be regarded as a kinematic one with an elastic element replacing a kinematic pair, occupy a half-way position. The elastic element is so much 'softer' than the spider that to a first approximation we can regard it as a rateless hinge, and so find the (angular) deflection in it. We can then multiply this deflection by the stiffness, and insert the moment so obtained to obtain a second approximation which is good enough for all practical purposes; a member which can be treated so is called *compliant*. In the treatment that follows the exact approach will be used.

The self-aligning planet of Figure 7.7 may be idealized as the system shown in Figure 7.10, which is a view along a radius through the centre P of the planet. The resultant of all the tooth forces could be taken as a single force F

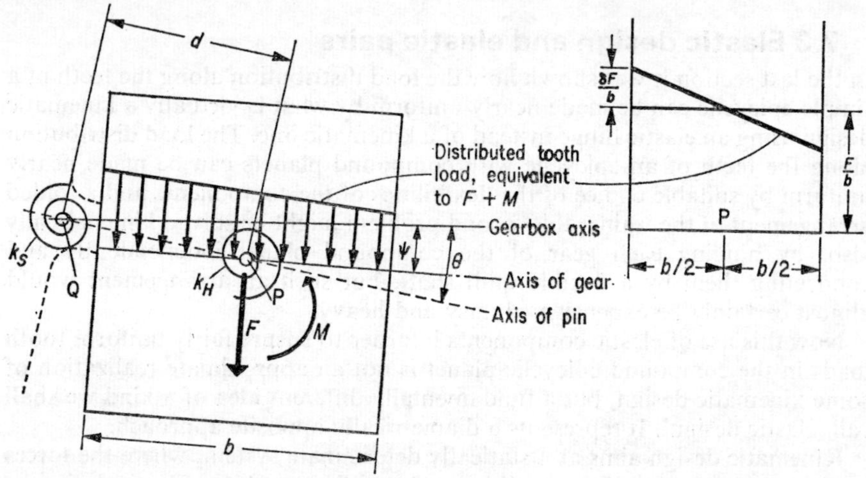

Figure 7.10 Idealization of Figure 7.7

acting through some point on the axis of the planet (since it is an idler). It is more convenient to reduce them to a force F acting through P and a moment M.

The elastic element, the diaphragm, can be regarded as a stiff hinge at P, of angular stiffness k_H, where angular stiffness is defined as restoring torque/angle turned through. The flexibility of the spider can be represented by another stiff hinge Q of angular stiffness k_S at a distance d from P; Q will be a little short of the end plate because of the bending in the pin itself.

Taking moments about P and Q,

$$\left.\begin{aligned} M &= k_H(\theta - \psi) \\ Fd - M &= k_S\theta \end{aligned}\right\} \tag{7.1}$$

where θ and ψ are the slopes of the pin and the planet respectively, as in Figure 7.10.

For the planet to have a slope ψ, it must force the annulus and sun teeth to slope. As these are generally very stiff, ψ will be small, and for a first approximation may be taken as zero. Eliminating θ from equations 7.1 gives

$$M = \frac{Fdk_H}{k_S + k_H} \simeq \frac{Fdk_H}{k_S} \tag{7.2}$$

Thus the essential requirement for good load distribution is that k_H should be small enough compared with k_S.

If the ratio of maximum to mean load per unit length is $1 + \delta$, as in Figure 7.10, the face width is b and the distribution of tooth load is linear, then it is easily shown that $\delta = 6dk_H/bk_S$, so that for a given value of δ which is regarded as acceptable, say, 0.08, we can calculate the required value of k_H/k_S. We then decide how small k_H can be made (the more space there is available, the smaller this will be) and hence how large k_S must be, i.e. how stiff a spider is required.

This seems like a rational design procedure, but the basic principle of not making arbitrary decisions (section 9.2) was violated when the flexible

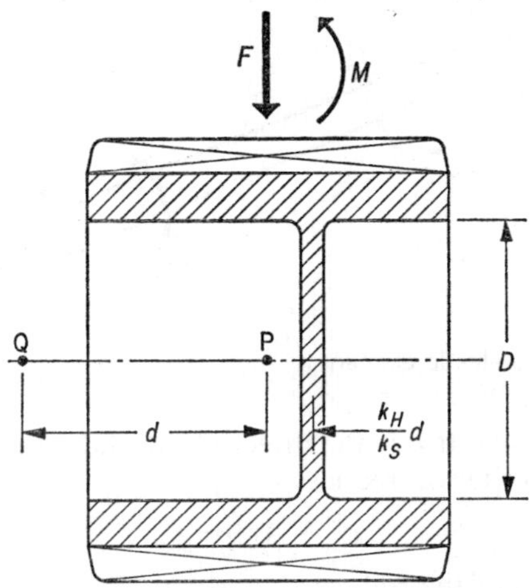

Figure 7.11 Redesigned geometry of Figure 7.7

diaphragm was placed in the middle plane of the planet. If it is put slightly further away from Q, so that the centre of the facewidth has a slight moment arm about it $(k_H/k_S)d$, as in Figure 7.11, then a uniform tooth loading will provide the necessary moment to align the teeth. If PQ is unaltered at d, then equations 7.1 become (with ψ = zero)

$$\left.\begin{aligned} M + F\frac{k_H}{k_S}d &= k_H\theta \\ Fd - M &= k_S\theta \end{aligned}\right\} \qquad (7.3)$$

and $M = 0$.

What has been gained by shifting the diaphragm? The facewidth of the planet can be reduced in the ratio $1 + \delta : 1$, and this is well worth while, but the problems of designing the diaphragm and the spider are scarcely altered, even though it may seem that the requirements for k_S and k_H are less stringent. To see this, it is necessary to look at the elements themselves.

The diaphragm has to withstand a shear load F and deflect through an angle θ; our improvement has not altered F, and has reduced θ only very slightly, and as these two quantities are a measure of the *task* to be performed the problem remains virtually unchanged.

As the diaphragm is difficult to analyse, another kind of elastic element will be used to illustrate the limitations of the tasks they can perform. Figure 7.12 shows an elastic hinge that is required to turn through an angle θ, as large as possible, while sustaining a given force F along its axis. It consists of a torsion rod AB of fixed length L encastered at A, made of material with allowable stress f on a basis of maximum shear stress (as in Figure 3.10: Tresca criterion)

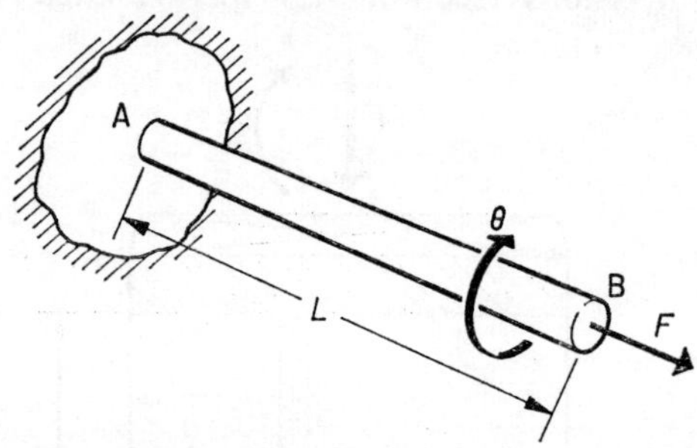

Figure 7.12 A simple elastic element

and a shear modulus G. If σ is the direct stress and τ the shear stress due to torsion, the Tresca criterion reduces to

$$\sigma^2 + 4\tau^2 = f^2 \tag{7.4}$$

and if r is the radius of the rod,

$$\sigma = \frac{F}{\pi r^2}, \ \tau = \frac{G\theta r}{L}. \tag{7.5}$$

Substituting in equation 7.4,

$$\frac{F^2}{\pi^2 r^4} + 4\frac{G^2\theta^2 r^2}{L^2} = f^2. \tag{7.6}$$

Differentiating with respect to r, and noting that, when θ is a maximum,

$$\frac{\mathrm{d}(\theta^2)}{\mathrm{d}r} = 0$$

$$\frac{-4F^2}{\pi^2 r^5} + 8\frac{G^2\theta^2 r}{L^2} = 0 \tag{7.7}$$

Eliminating r from equations 7.6 and 7.7 yields

$$\theta = 0.55\frac{f^{3/2}L}{GF^{1/2}} \tag{7.8}$$

This is the maximum value of θ that can be had under the prescribed conditions. The requirement to support a load F means that the value of θ obtainable with the given geometry and in the space available is limited.

A similar limit exists in the case of the diaphragm of Figures 7.7 and 7.11. It occupies a limited space (diameter D in Figure 7.11) and carries a shear force F, so that the value of θ is limited. From equations 7.3,

$$\theta = \frac{Fd}{k_S}$$

and this dictates the required value of spider stiffness k_S. For the central diaphragm

$$\theta = \frac{Fd}{k_S + k_H}$$

but the reduction in facewidth has reduced d, so the off-centre diaphragm is a slight help.

The result given in equation 7.8 depends on the geometry of the elastic element. If we used a spaced bundle of round rods, or a stack of flat leaves, a greater value of θ could be had for the same F and L. The task consists of a force term (F) and a displacement term (θ). If these are directly coupled, i.e. if the force is in the direction of the displacement, the element is a spring and is governed by the simple limits noted in section 4.5. But in the more typical cases of Figure 7.13, the writer knows of no absolute limits on the task (F, θ) that can be performed by an element of given material occupying a given space. Nevertheless, it seems sure that such limits must exist, and they certainly do as far as practical designs are concerned. Unfortunately, it is often difficult to find even the practical limits applying to specific geometries—the torsion rod is not typical in this respect. Examples are given among the questions at the end of this chapter.

Sometimes the task we want the element to perform demands a low stiffness as well as or instead of a high allowable displacement. That these requirements are different can readily be seen in the case of the torsion rod. Equation 7.6 may be regarded as showing the 'Tresca strength-squared' being shared between direct and torsional loadings, in a ratio which proves to be 1:2. For a very small θ we can share the 'strength-squared' almost in the ratio 3:0, the radius can be reduced by a factor of $(1/3)^{1/4}$ and the stiffness to 1/3 of the value for the maximum θ rod. The diaphragm for the planet gear of Figure 7.7 must be designed for maximum θ, rather than low stiffness.

Before leaving this example, let us look quickly at the effect that manufacturing tolerances have on load distribution. The effect of helical errors of the same hand and amount ψ (see Figure 7.10) in the annulus and sun of the epicyclic gear arranged as in Figure 7.11 is to introduce a moment M of roughly ψk_H. With a kinematic hinge, $k_H = 0$ and so $M = 0$, and the manufacturing tolerances have no effect on the load distribution. The tolerances also increase the task for which the elastic element must be designed.

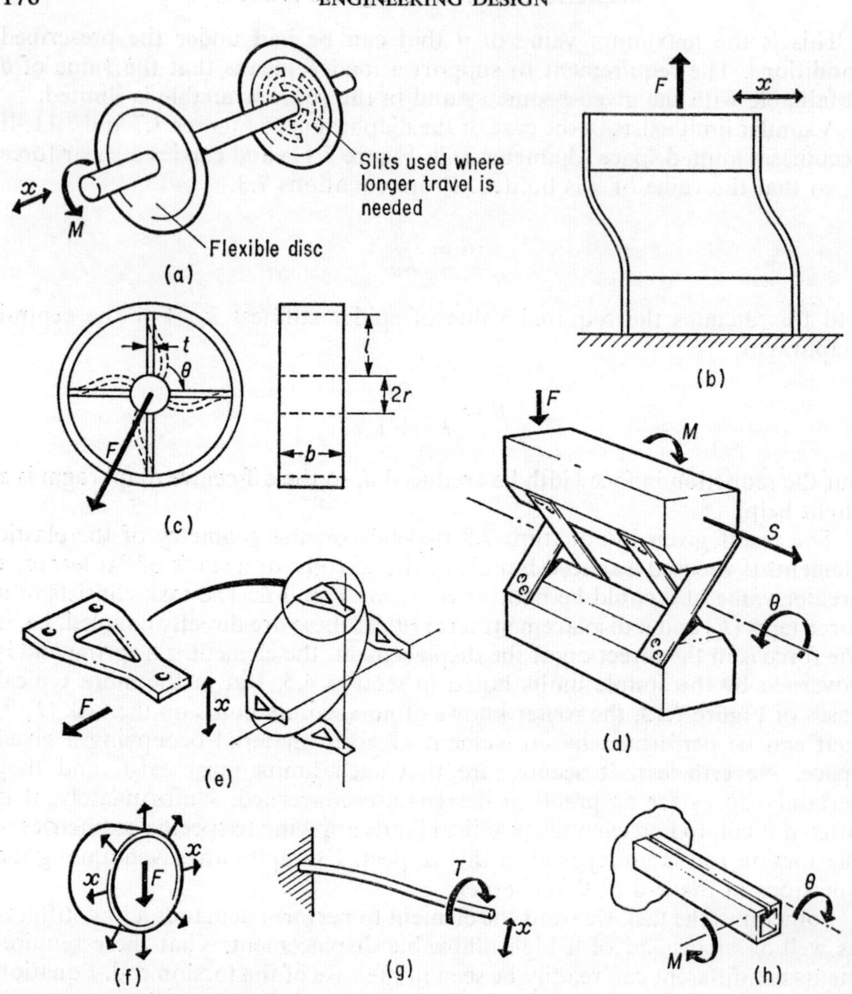

Figure 7.13 Elastic elements

The three arrangements for distributing the load on large tilting pad thrust bearings shown in Figure 7.2 are

(a) a pure kinematic design,
(b) a hybrid, and
(c) an elastic design.

The three kinds of load balancing possible in such arrangements are:

(1) equalizing of loads between pad and pad,
(2) location of the centres of pressure on individual pads, and
(3) more detailed control of the load distribution on individual pads than is implied by (2).

Only Figure 7.2(c), with a multiplicity of supporting springs under a relatively flexible pad, can provide (3). Arrangement (a) provides (1) and (2) by purely kinematic means, while (b) provides (2) largely by kinematic means and a measure of (1) by elastic means. The springs used in (3) are heavily preloaded by central telescopic ties which collapse in compression. The task performed by these springs is measured by the force they must exert and their softness (i.e. the reciprocal of their stiffness), which is necessary to allow for manufacturing tolerances without too much variation of loading. Because the force F and the stiffness k are in the same direction, there is an absolute lower limit to the amount of material needed for a linear spring to perform this task (see section 4.5).

Figure 7.13 is effectively a page of the engineering repertoire, showing a selection of the commoner elastic elements. The loads shown are not the only ones the elements can sustain, but have been chosen to show possible uses.

The elastic equivalents of kinematic slides, Figure 7.13(a) and (b) are frequently used, as is the 'crossed-spring pivot' (d) widely met in instruments such as wind-tunnel balances. Figure 7.13(c) is a more expensive form of crossed-spring pivot.

One of the most useful of all elastic elements is the cantilevered spring (e)—three or four arranged as in the small inset are useful for dealing with the centring problem discussed in the next section, and two can be used in arrangement (b). The short thin walled tube (f) will serve a similar purpose, since it can be stretched radially relatively easily while being very stiff in bending and shear.

The elements shown in Figure 7.13(g) and (h) are complementary in function—the first transmits a torque while permitting small transverse movements, while the second is easily twisted but stiff in bending.

7.4 Comparison of kinematic and elastic means

There are a few other, rare, means of guiding or locating one body relative to another, such as electromagnetic induction; we will not include hydrodynamic and hydrostatic bearings, which are effectively kinematic, so short is the distance between the mating surfaces. In the majority of cases the only practical choice is a kinematic means, but there are many interesting problems where both kinematic and elastic means are worth study.

Advantages of kinematic devices

(1) Smaller when 'displacement term' is not very small.
(2) Free of stiffness (zero rate) in desired direction of movement.
(3) Available as standard items.
(4) High stiffness in 'fixed' directions.
(5) Load-balancing functions not affected by tolerances.

Disadvantages of kinematic devices

(1) Larger when displacement is very small.
(2) Subject to hysteresis *or* dependent upon pressurized lubrication *or* requiring to be kept moving *or* must be of rolling type.

(3) Subject to backlash.
(4) Difficult to protect from corrosion and dirt, and sometimes 'brinelling'.
(5) Requiring maintenance.

Advantages of elastic devices

(1) Smaller when displacement is very small and a high ratio of stiffness in 'fixed' and displacement directions is not required.
(2) Free of hysteresis and backlash.
(3) Requiring no maintenance.
(4) 'Cut-price' solutions are sometimes possible.

Disadvantages of elastic devices

(1) Larger when displacement is large or stiffness must be low, or ratio of stiffnesses must be large.
(2) Stiff in displacement direction.
(3) Sometimes expensive.
(4) Careful and expensive design sometimes needed to avoid fretting.
(5) Load balancing upset by tolerances.

A few cases will be discussed.

Further discussion of load sharing in gears

It is economical in gearing with high maximum torques to use multiple layshafts, and the planets and compound planets of epicyclic gears are, in effect, multiple layshafts. It has already been shown that load sharing between the layshafts in simple epicyclics can be achieved kinematically by floating sun or annulus, or elastically by flexibility in the annulus. With more than three planets, the elastic method still works but the kinematic does not (section 8.2).

With multiple layshafts that are not epicyclic in action, a common number is two. Load sharing here is generally done by flexible quill-shafts between layshaft pinions and wheels, but kinematic means have been used, e.g. by end thrust balancing with helical gearing. Notice that the flexible layshafts represent a true elastic design, not an approximation to a kinematic design.

Centring devices

The bores of turbine and compressor discs may grow relative to their shafts due to centrifugal and thermal stresses by as much as 0.003 times their diameter.

Figure 7.14(a) to (d) show ways of holding the discs concentric with the shafts under these circumstances. Diagrams (a) and (b) show kinematic means, which require very high standards of accuracy in manufacture; the face-serrations (b) are ill-adapted to the centring of long stacks of discs on a separate shaft. Cheap versions of (b) may be had by inserting radial dowels with their axes lying in the plane of two butted faces.

The simplest elastic solution is a large interference. Where this is not acceptable, the arrangement shown in Figure 7.14(c) may be possible. A similar solution is shown in Figure 2.4 (Rolls-Royce Ltd.). The serrated ribs

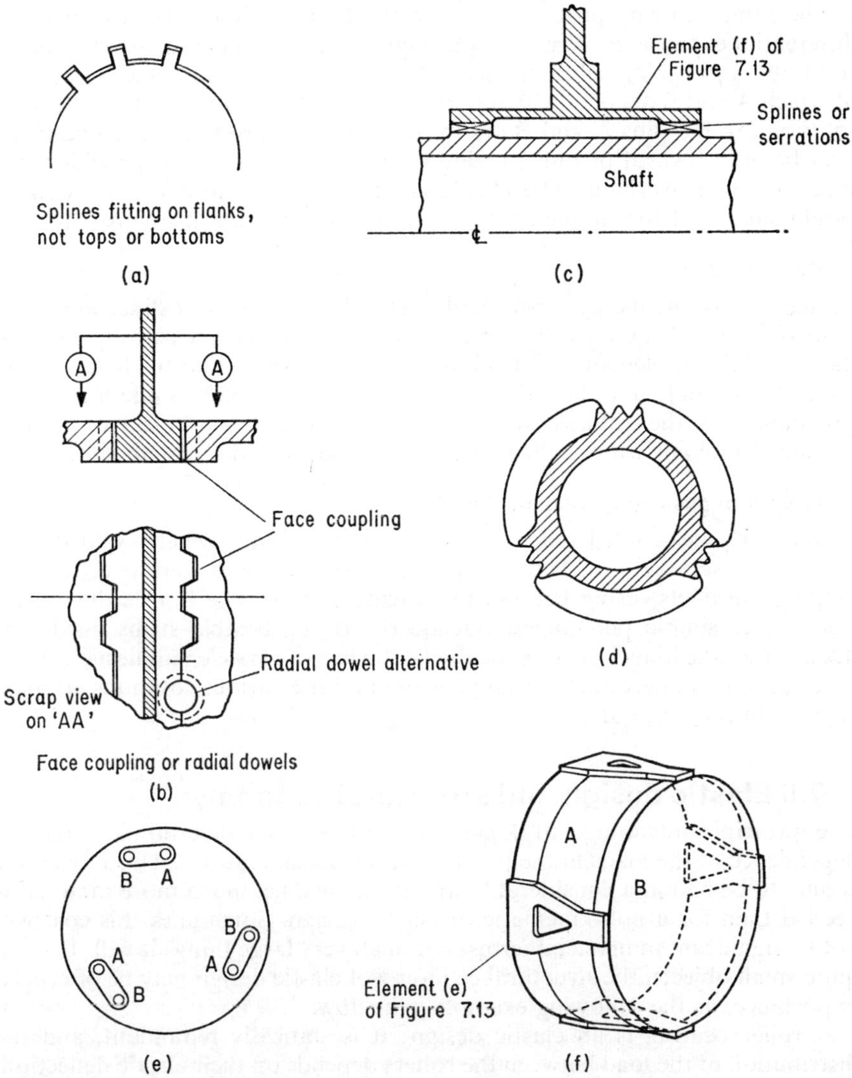

Figure 7.14 Centring devices

in (d), an unproven solution systematically invented by students under the writer's guidance, provide local centrifugal loads which bend the shaft wall so that the small initial interference is not lost. The writer believes that this may be the best solution where the shaft wall thickness is small, but where the disposition problem of section 6.3 arises it is difficult to find an embodiment which is both light and simple to make. Because the shaft wall is bent, and not stretched, ribs of quite small section suffice.

The same centring problem arises with stationary components owing to differential expansions. Freed of the high centrifugal field, a wider range of solutions is possible, including those shown in Figure 7.14(e) and (f), in both of which A and B denote the two parts to be kept concentric. In the kinematic solution (e), the pins A and B are in parts A and B, respectively. Eccentrics may be used instead of links, giving a compact embodiment, provided it is sure they will overhaul. The elastic solution (f) uses four elements (three would suffice) of the simple cantilever spring sort shown in Figure 7.13(e).

Instruments

In many instruments, hysteresis and backlash in pivots and slides are much more objectionable than a slight rate, and circumstances often strongly favour the use of elastic elements. Often both the displacement and the load term in the task are small, and the stiffness can also be made small; sometimes, as in galvanometers, the stiffness can be made use of (or a cross-stiffness, as in some twisted strip comparators which also use an elastic slide—Johansson).

Hinges in gas-cooled reactor ducts

In the early gas-cooled reactors for power generation, where welded steel pressure vessels were used, thermal-expansion difficulties were overcome by hinging the ducts, using bellows to provide sealing. The hinges themselves were either simple pin hinges, outside the ducts, flexible strips inside the ducts, or elastic hinges outside the ducts, the last of very elegant design. A full account of this interesting design problem and the various solutions adopted is given in reference 51.

7.5 Elastic design and structural economy

The principle advantage of kinematic design is its determinacy, the independence of the loads in the parts of elastic considerations. This advantage is only to be had at a considerable structural cost, i.e. much more material is needed than for a non-kinematic or elastic design. Sometimes this cost will not be significant in financial terms, but in all very large things it will. Even in quite small objects, the structural economy of elastic design may be of crucial importance, as the following example will show.

A roller bearing is an elastic design; it is statically redundant, and the distribution of the load between the rollers depends on their elastic deflection. We could make a kinematic roller bearing with three rollers; it would be about seven times the volume and would have several other very undesirable attributes. Notice in passing that we can improve the load sharing (or increase the tolerances on the components) in a roller bearing if the elasticity is increased, e.g. by making a hole through the middle of the elements. This has been done both in parallel and taper roller bearings. Notice also that the lengthwise distribution of load on a cylindrical roller on a cylindrical race will have peaks at the ends because of variation of the stiffness due to three-dimensional effects. This can be corrected by the last of the three means given in section 7.2, by making the rollers not truly cylindrical but slightly relieved or 'blended' at the ends.

The helicopter engine in Figure 7.1 is a massive object traversing a clear space in the fuselage; it seems a waste that it cannot be used as a structural component of the airframe. Apart from the extra difficulty set by the large thermal expansion of this potential member, this is a degree of sophistication not yet reached in aircraft design.

Consider a very simple structure, a uniform horizontal beam AB, subject to a single vertical load W. Figure 7.15 shows two versions, (a) pinned at both ends, i.e. kinematic (except for the redundant constraint in the AB direction, which need not concern us) and (b) a redundant version encastered at both

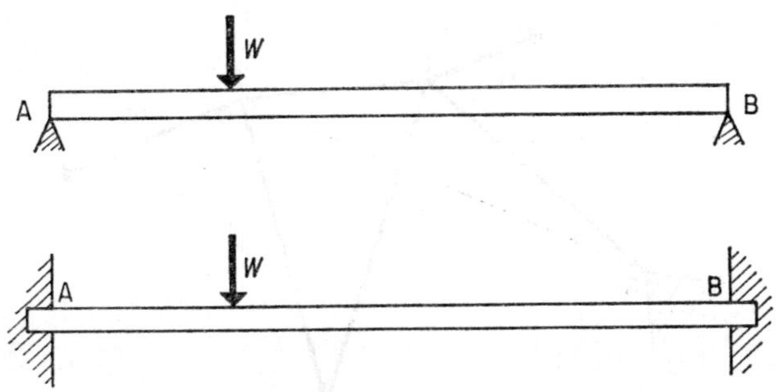

Figure 7.15

ends. Wherever W acts, the bending moment is less in the encastered beam because it is receiving more help from its supports, which in a mechanical-engineering application will generally be further parts of the structure. Even a pin-jointed frame with given nodes, if subject to different systems of loads at different times, will probably take less material if it is redundant [17]. Suppose such a frame is fixed at A and B, and loads W_1 and W_2 are to be applied at C and D (Figure 7.16). Then if a frame is provided to support W_1 at C from A and B, part of the load W_2 can be routed via C. For the particular configuration shown, it looks very likely that it will be more economical to route as much as possible via C. Depending on the ratio of W_1 and W_2, it may not be possible to take all the load via C, and then rather than increase the scantlings of the members AC and BC, say, it may be cheaper to go straight to A. We now have a redundant structure, and some of W_1 will go by D, so that AC and BC should perhaps be thinned—the problem is very complicated, but the key point is this: the route to 'earth' from D is cheaper via C *only* when part of that route, from C to earth, is an existing *asset* (see section 8.1).

Note also that in the frame of Figure 7.16 it might pay to have extra nodes between CD and AB, depending on end costs and the structure loading coefficient. The structure loading coefficient, already mentioned in section 6.7, is a parameter expressing the size of the loads relative to the size of the structure and the strength of its material. In this case we might use the dimensionless form

$$\frac{W_1}{(AB)^2 f}$$

where f is the allowable stress. If the structure loading coefficient is low, the scantlings are small compared with AB, the members are slender and so buckling is an important consideration, favouring extra nodes in this case, and more elaborate structures in general.

Another way of seeing the structural superiority of elastic design is via the idea of *load diffusion*. If we wish to locate one body X relative to another Y by a kinematic design, all the loads in X must be gathered up into a relatively

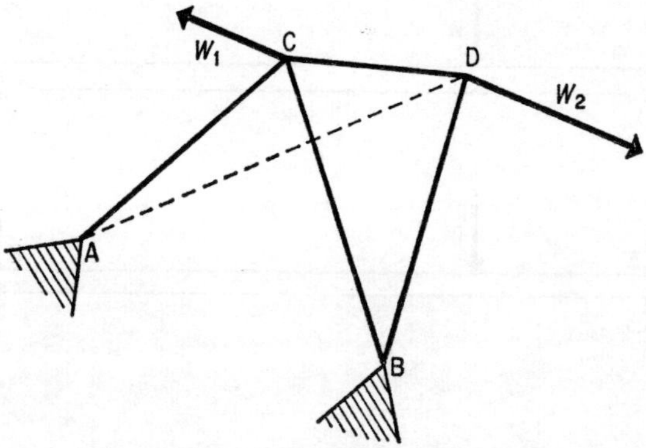

Figure 7.16

few relatively small zones which are attached to Y, and then diffused out again all over Y. If we look outwards, as it were, from the attachment points, we have two *load diffusion* problems. The cost of load diffusion will be high when the structural loading coefficient is low. A good example is balloon design, where the buoyant forces spread over a vast area of flimsy envelope must be gathered up to support a small dense load. A parallel is the supporting of the spherical propellent tank in the ELDO rocket third stage by 192 titanium alloy straps—notice the structure loading coefficient here was low largely because of a very high allowable stress.

Where the structure loading coefficient is high, the cost of gathering up loads is much less; however, once gathered it may be worth keeping them so, as in those diesel engines that use stiff-jointed frames for their structure [33]. Non-structural casings close in the openings in these frames, a case of separation of function.

With the tendency of engineering devices to become bigger and to use smaller quantities of stronger but more expensive materials, the relative number of cases is increasing where the structural economy of elastic design is more valuable than the determinacy of kinematic design.

Questions

Q.7.1(2). Figure 7.17 shows a section commonly used for the slideways of lathes (only the saddle ways are shown). Alignment tests on the accuracy of

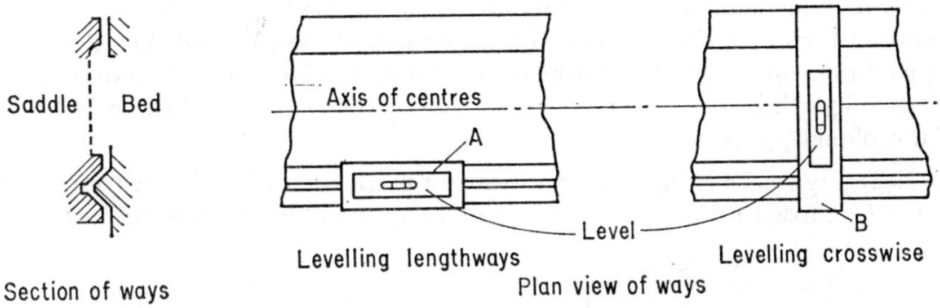

Figure 7.17 Alignment tests on lathe

lathes call for checks to be made of the levelness of the front (inverted V) and back (flat) ways separately, and also of the truth of front to back ways by means of a bridge piece straddling them, a sort of dummy saddle. Think about this with regard to (a) the value of such tests (b) the importance of the different kinds of error which might be found and (c) the nature of the pieces A and B (Figure 7.17) used to adapt the sensitive level to these measurements.

Q.7.2(2). Figure 7.18 shows a duct subject to an internal pressure p, at a point where an elastic hinge is provided. The continuity of the pressure wall is

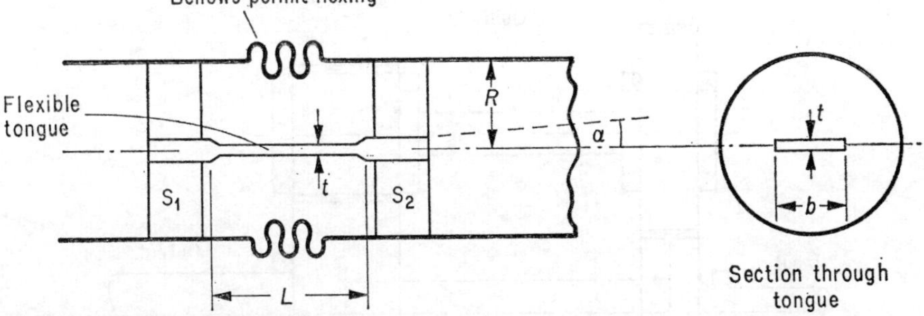

Figure 7.18 Virtual hinge in duct

maintained by a bellows which is relieved of the axial pressure load $p\pi R^2$ by a flexible tongue of effective length L anchored in two spiders S_1, S_2. This tongue and the bellows permit flexure to an angle α from the straight, necessary to accommodate differential thermal expansions in the system of which it forms a part. Estimate very roughly the necessary length L, assuming the hinge is free from shear forces, or axial forces in excess of that due to internal pressure. The allowable stress in the material of the tongue is f and the Young's modulus is E.

Q.7.3(2). Consider the design of the flexible pin in the self-aligning pinion arrangement of Figure 7.8, in particular the optimum variation of section along its length. Do not plunge into a flurry of mathematics, but see first how far common sense will take you. For example, is there any portion of the length in which an infinite EI would be desirable? Determine a practical optimum form, bearing in mind the stubbiness of the pin, and then derive an approximate attainable θ in terms of the notation of section 7.3, assuming the effective length of the pin is $2d$ symmetrically disposed about the mid length of the planet pinion.

Q.7.4(2). Figure 7.19 shows a three-bearing shaft assembly, with bearings at A, B, and C. Close to B there is a gear tooth coupling of base circle radius b, with

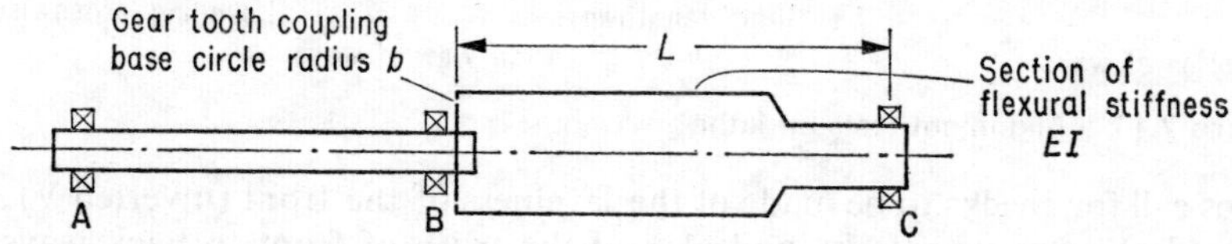

Figure 7.19 Three-bearing shaft

a coefficient of friction μ between the teeth. The portion AB of the shaft may be regarded as stiff, and the portion BC, of length L, has a flexural stiffness EI. The torque in the shaft at the gear tooth coupling is T. Estimate (a) the axial force the gear tooth coupling can sustain without slipping and (b) the amount δ that bearing C must lie off the axis of AB for the misalignment to cause slipping in the gear-tooth coupling (now assumed free of axial load).

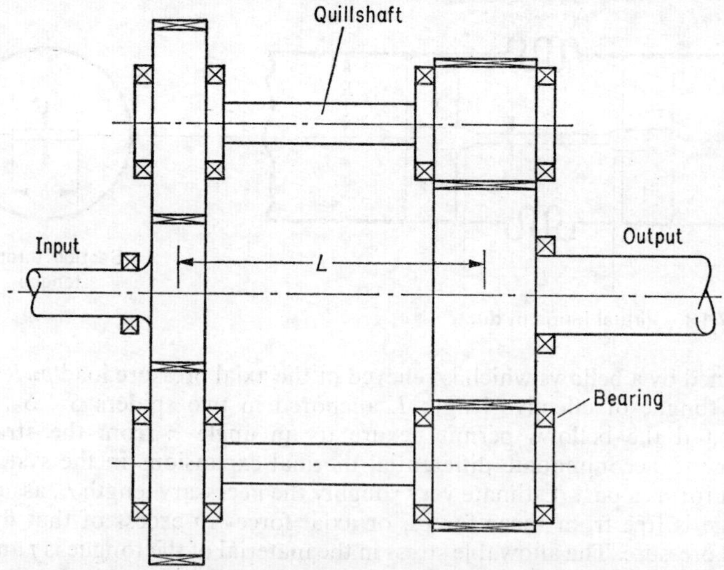

Figure 7.20 Twin layshaft reduction gear

Q.7.5(3). Consider the problem of ensuring equal sharing of the loads in a simple epicyclic with n planets, where $n > 3$. How will the efficacy of a flexible annulus gear vary with n? What other means are available?

Q.7.6(1). A twin layshaft reduction gear of the layout shown in Figure 7.20 is such that the sum of the torques in the two layshafts is $2T$; if all the elements concerned are at the appropriate extremes of their tolerances, and a light torque is applied to the input shaft, the lagging layshaft pinion has to be advanced an angle ε to make contact on the driving flanks.

Using the allowable stress diagram of Figure 3.10, find the shortest distance L with which it is possible to ensure that the torque distribution is not more unbalanced than $1.1T$ to $0.9T$.

Answers

A.7.1(a). The writer doubts the value of such tests, since all that is of practical importance is the truth of the saddle motion, which should be measured directly. It can be argued, however, that a fortuitously or deliberately produced arrangement of high spots could give a true saddle motion on irregular

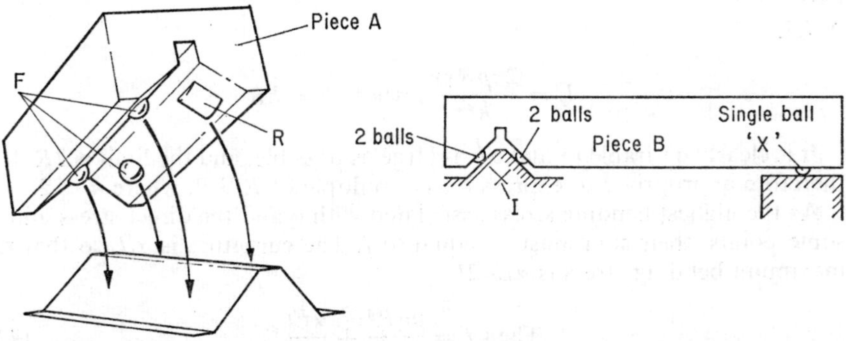

Figure 7.21 Alignment tests on lathe

ways; the high spots would soon wear, and the quality of the lathe would rapidly deteriorate. In effect, this is an argument that the tests are necessary to check the quality of the guiding surfaces, in which case it would surely be better to use tests specifically designed for the purpose.

(b) If the front and back ways rise and fall together, the tool merely rises and falls and there is no practical effect on the accuracy of the work. If, however, the ways cross-wind or warp, the front rising when the back falls and vice versa, the saddle will roll and the tool move in and out of the work.

(c) The pieces A and B should both be constrained to have only one degree of freedom, i.e. translation along the ways. Accordingly five of their potential six degrees of freedom must be suppressed by a total of five simple constraints. For piece A these can be provided by three balls F (see Figure 7.21) and a roller R, the former providing three simple constraints, and the latter two simple constraints, since a line is constrained to lie in a plane.

Piece B is a little more difficult. At least one constraint must be associated with the back way, otherwise the motion will be independent of it. There cannot be as many as three associated with the back way, since it is a nominal plane and with three constraints would limit the motions of B to two translations and rotation about a vertical axis; the important displacement to detect, however, is rotation about an axis parallel to the headstock, as has been discussed under (b). There must be one or two simple constraints on the back way.

With one simple constraint at the back way, we need four at the front, and this can be achieved by substituting another ball for the roller in piece A (Figure 7.21). Piece B is then free to rotate about the instantaneous axis I associated with the four balls on the front way until ball **X** rests on the back way.

With two simple constraints on the back way, say, two spaced balls at X, then only three balls must be used on the front way, and an additional closing force must be used to keep the third ball in contact with the bed, just as a T-square must be kept pressed against the edge of the drawing board. Thus the former arrangement, with only one ball on the back way, is to be preferred.

A.7.2.

$$L = \frac{2\pi p R \alpha E}{k f^2}, \text{ where } b = kR$$

It is clearly desirable to make b as large as possible, and the limit is $2R$. It is therefore appropriate to express b as a multiple of R, kR, where $k < 2$.

As the highest bending stress associated with α and the direct stress add at some points, their sum must be equal to f. The curvature is α/L so that the maximum bending stress is $\alpha Et/2L$

$$\text{Then } f = \frac{p\pi R^2}{kRt} + \frac{\alpha Et}{2L} \tag{7.9}$$

Since f is fixed, equation 7.9 can be differentiated with respect to t, giving

$$0 = -\frac{p\pi R}{kt^2} + \frac{\alpha E}{2L} - \frac{\alpha Et}{2L^2}\frac{\mathrm{d}L}{\mathrm{d}t} \tag{7.10}$$

and when L is a minimum $\mathrm{d}L/\mathrm{d}t = 0$. Either by this route, or simply by recognizing the equal moduli of the powers of t in equation 7.9, it is found that L is a minimum when f is equally shared between bending and direct stress (see section 6.10) and hence the required answer can be obtained.

In practice, it would be necessary to allow for shear loads, S, on the hinge, and also for secondary bending moments due to the deflection. The bending stress due to the shear load increases with L, thus imposing a limit on what can be done in the way of reducing stresses by making the tongue longer, and the task represented by ($p\pi R^2$, S, α) may be beyond any available material. It would then be necessary to adopt some other solution [51].

Notice that in this problem a figure of merit for the material is f^2/E, which is also a figure of merit for a spring material (section 4.5). Note also that by 'building bent' it might in some cases be contrived that the bending stresses alone were present when cold and depressurized and the direct stresses alone when hot, or some variation on this theme.

A.7.3. Since only curvature which produces a slope opposed to θ is helpful, the portion of the pin from the line of action of F to the spider should be as stiff as possible. The remainder should be fully stressed, which means that it should be thinnest on the line of action of F, and increased in diameter to the right (Figure 7.8), in order to withstand the increasing bending moment. Allowing for the stubbiness of the pin, however, it is doubtful whether it is practicable to exploit this 'ideal' form, and a good enough approximation is probably achieved with two diameters, as large as possible to the left of F and as small as possible to the right.

The moment at the free end of the pin is Fd, so that the radius r of the thin part of the pin is given by

$$f = Fd/\tfrac{1}{4}\pi r^3 \qquad (7.11)$$

The slope developed in the length d of the thin part is

$$\frac{Fd^2}{2EI} = \frac{2Fd^2}{\pi E r^4} \qquad (7.12)$$

Putting this slope equal to θ and using equation 7.11 gives

$$\theta = \frac{df}{2E}\left(\frac{\pi f}{4Fd}\right)^{\frac{1}{3}}$$

A.7.4. (a) $\mu T/b$. (b) Of the order

$$\frac{2\mu T p L^2}{3\pi b E I}$$

where p is the pitch circle radius.

The loads on the teeth total T/b, so the force required to slide them axially is $\mu T/b$. If the total tooth load T/b is distributed uniformly round the gear tooth coupling, then to 'bend' it about the horizontal diameter, say, requires all the teeth in the top half to slide one way and all the teeth in the bottom half the other. If p is the pitch circle radius, the moment arm of each set of forces about the diameter is $(2/\pi)p$, so that the resisting moment is

$$2 \times \frac{\mu T}{2b} \times \frac{2}{\pi} p \text{ (see Figure 7.22)}$$

If the shaft BC is regarded as encastered at B and given a deflection δ at C by a load F,

$$\delta = \frac{FL^3}{3EI}$$

and the moment at B is

$$FL = \frac{2\mu Tp}{\pi b}$$

Once the coupling slips, however, and takes on an angle, the skewing of the teeth relative to one another will concentrate the load in two arcs near the ends of the horizontal diameter, greatly reducing the resisting torque.

These rough calculations, with values substituted, will indicate whether it is safe to rely on a gear tooth coupling or serration to allow axial sliding. The writer made an experiment on a running gas turbine with the configuration of

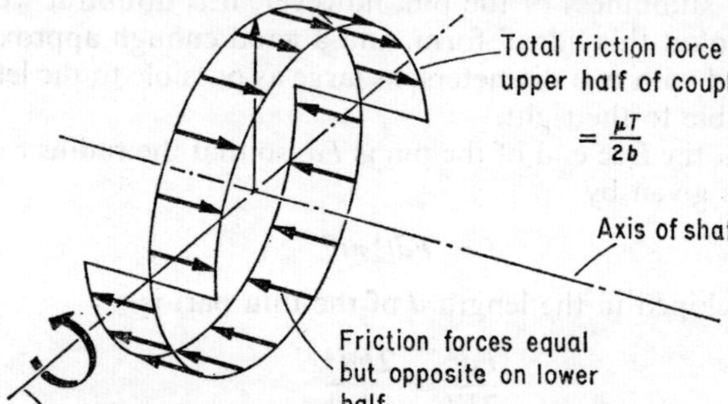

Figure 7.22

Figure 7.19 and found that the force needed to slide the coupling corresponded to three times the capacity of the separate thrust bearings on the shafts AB and BC; this would agree with the expression $\mu T/b$ with a high value of μ (about 0.5). The hope that gradual slip would occur because of wriggling of the teeth by misalignment of the bearings proved illfounded, as might have been shown by the sort of calculation given above. The δ required was larger than would normally occur.

A.7.5. A full analysis of this problem will not be given, but here are a few notes. The load-sharing properties of a flexible annulus gear depend on the tooth load at any mesh causing a deflection which reduces the 'interference' as it were, at that mesh. There are two such flexibilities, one associated with the radial component and the other with the tangential component of tooth load. The effect of the former may be likened to that of an elastic band round a nominally symmetrical bundle of pencils, where even if the pencils vary slightly in diameter the force exerted on them by the band is nearly the same (notice, however, the band works in tension and the gear in bending).

A study will show that the stiffness to the radial component approaches proportionality to n^2: while it may well be low enough at $n = 4$ to effect load sharing, at $n = 8$ it will be about four times as great and so much less useful.

The other flexibility arises from the moment arm of the tangential component of tooth load about the principal axis of the ring cross-section; the ring bends into a wavy form with n wavelengths in the circumference. This form of flexibility decreases only as n^{-1}, but is smaller in the first place, e.g. at $n = 4$.

Thus it is clear that as n increases, the flexibility of the annulus decreases in efficacy as a load-sharing agency. The point at which it becomes inadequate will depend on the accuracy of the gears, the level of stressing, and the fineness of the teeth.

When annulus gear flexibility is not enough, springs between the separate planets and the spider can be used, e.g. the arrangement of Figure 7.8 will serve this purpose as well as aligning the planets, and in this respect it is superior to that of Figure 7.7. Alternatively, hydraulic means between the planets and the spider might conceivably be used, as might kinematic means.

A.7.6.

$$12.3\,\frac{GT^{\frac{1}{3}}\varepsilon}{f^{4/3}} \text{ where } G \text{ is the shear modulus.}$$

The difference between $1.1T$ and $0.9T$ must produce a wind-up ε in a quill shaft. The quill shaft must also be able to withstand $1.1T$. To be as flexible as possible for a given strength it must be solid. If the quill shaft radius is r, then

$$\varepsilon = \frac{0.2TL}{\frac{1}{2}\pi r^4 G}, \quad \frac{f}{2} = \frac{1.1T}{\frac{1}{2}\pi r^3}$$

Eliminating r gives the answer. Notice that the figure of merit here is $f^{4/3}/G$, as in the elastic element of Question 7.3 except for a different modulus. Note also that the form $f^{3/2}/G$ arises in the study of the element of section 7.3 (equation 7.8) and f^2/E in the element of Question 7.2.

8
Tasks, Costs, Assets, and Values

8.1 Introduction

The category of ideas dealt with in this chapter may well prove very important in the long run. Unfortunately, it is as yet little developed. Value engineering is a related idea, concerned with the embodiment rather than the conceptual stage, and less fundamental in its approach.

Tasks

Several instances have been given in this book of the quantification of a function to be performed; the term *task* is used to denote a quantified function. Section 1.9 introduced the idea of the quantification of the resistant function of a pressure vessel as holding-together power, and in section 2.5 the possibility of quantifying the structural functions in an L.N.G. tanker was mentioned. Neither in the case of the tanker nor that of the beam mentioned in section 4.6 was the quantification carried out—it is fairly easy for the former and difficult and uncertain for the latter [17]. In section 4.5 the liquefaction of nitrogen was regarded as the task of increasing its availability function, b, by a certain amount. In section 7.3 the task of an elastic element was stated in terms of a force F and displacement θ, but the two were not combined into a single quantity. Once we can quantify functions, we can often estimate the costs of performing them. For example, for steel structures erected on site (shipyards produce more cheaply) we may for £1 get up to 160 kilojoules (1.4×10^6 lbf in.) of holding-together power if we can work at the favourable point A of Figure 3.10, or 80 kJ (0.7×10^6 lbf in.) at the less favourable but much commoner working point B.

In most practical structures we cannot even hope to work all the material at B, and an efficiency of use will have to be guessed (see section 4.5). Nevertheless, we can estimate likely task/cost ratios closely enough for many of the purposes of conceptual design.

Values

The ability to perform a task we call a value, but it is convenient to use the term rather widely. For instance, in disposition problems we saw how different portions of a space to be disposed of had different values for different purposes—the material they could accommodate would be able to perform different amounts of the various tasks in hand. In the simple problem of the splined shaft with turbine discs (section 6.3), bits of space in the annular region r_1 to r_0 occupied by the splines had great value for shaft stiffening but none for the disc-holding-together task.

Assets

An asset is an existing value provided for some other purpose. Thus, in the frame sketched in Figure 7.16, the structure used to support the load W_1 at C is an asset when we come to design structure to support a load W_2 at D (W_1 and W_2 act at different times).

The asset outlook is specially valuable when a scheme has reluctantly been increased in complexity. It is natural for the designer to resist additions to his first concept that are forced upon him by more detailed study. But once it has been accepted that such an addition is essential, he should then try to draw what further advantage he can from it. In comparing alternative schemes it is easy to remark that one involves some extra component, but not so easy to make sure that that component is properly regarded as an asset and exploited to the full. Also, some devices have inherent in them assets which cannot ordinarily be turned to full account, but may be exploited fully in special circumstances (see section 8.3).

8.2 Tasks and values in gearing

A good simple measure of the task performed by reduction gearing is the output torque. In Figure 8.1 the cost per unit of output torque is plotted against reduction ratio for data from two sources [34]. The near constancy of the cost per unit of torque shows that it is an apt measure of the task.

To see why this should be so, refer back to the scaling study of gears in section 4.7, where it was shown that the allowable torque T_p in a pinion was given roughly by

$$T_p = \frac{CD_p{}^2L}{1+k} \qquad (8.1)$$

where C is a constant, D_p the pinion pitch diameter, L the facewidth and k the ratio of the number of teeth in the pinion to the number of teeth in the gear ($k < 1$).

If T_0 is the output torque, then $T_p = kT_0$. Equation 8.1 gives

$$D_p{}^2L = \frac{(1+k)kT_0}{C}$$

and the volume of the pinion is a fair measure of its cost. As the teeth of the output gear are similar to those of the pinion, it is reasonable to assume that its cost is about $1/k$ times as much, and so we may expect the cost of the output train to be roughly proportional to

$$\left(1+\frac{1}{k}\right)(1+k)\,kT_0 = (1+k)^2T_0$$

Not much weight is to be placed on this particular result—there should be several terms allowing for different parts of the cost—but the proportionality to T_0 and the slow variation with k should be noted. When the other terms are added, there is a reversal of the steady reduction of expression 8.2 at small enough k's. Small k's will mean very large diameter gears that cannot be costed simply per yard of rim, as it were, and the cases to hold them will also increase rapidly in size [35]. The effective cost of the train must start to rise at

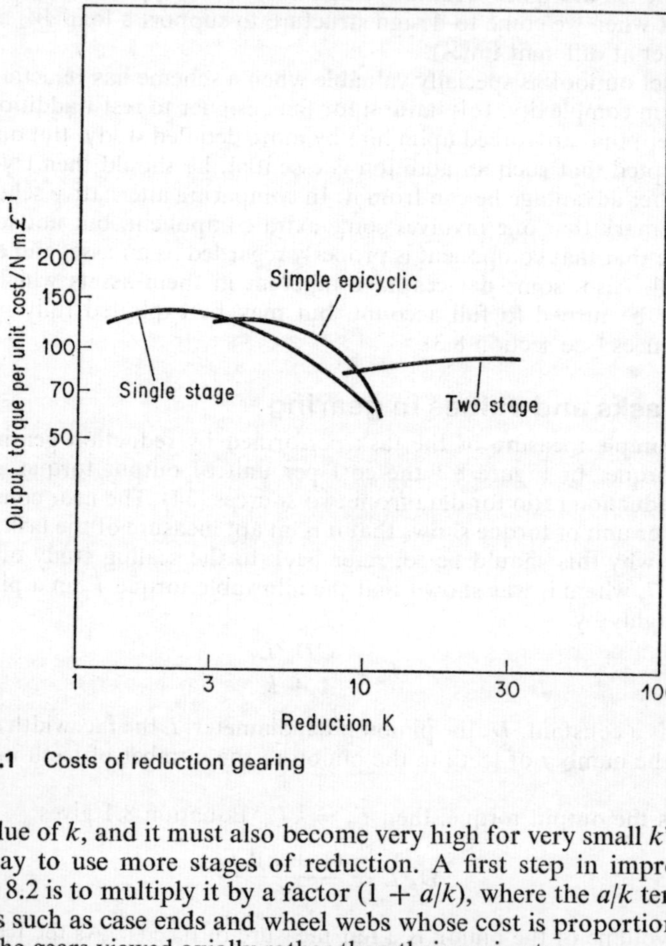

Figure 8.1 Costs of reduction gearing

some value of k, and it must also become very high for very small k's, so that it will pay to use more stages of reduction. A first step in improving expression 8.2 is to multiply it by a factor $(1 + a/k)$, where the a/k term allows for items such as case ends and wheel webs whose cost is proportional to the area of the gears viewed axially rather than their perimeter.

We can expect a minimum cost for a train of given output torque at some value of k, k_M say. When the overall reduction ratio K ($K < 1$) is sufficiently small to require two stages, a simple optimization will divide the ratio between them in such a way that the output stage has a ratio closer to k_M than that of the input stage, since it is larger and more expensive and a fractional increase in its cost is more significant. However, there is a large saving in making the centre distance on both stages the same, so that the bores for the input and output shafts are concentric. With this extra constraint and an empirical value for a, the balance of facewidths is determined. Lower values of k must be expected to give a somewhat higher cost per unit of output torque because they lead, first to uneconomical values of k in the stages, and eventually to the adoption of an extra stage with a net increase of cost despite the return to economical k's.

The cost of reduction gearing is closely associated with the output torque, largely because of the need for a massive final-stage wheel to carry that torque. The most promising way of reducing cost would seem to be to reduce the size and cost of this 'bull wheel'.

In order to do this we might:

(a) use the potential value of the teeth more fully, by better load distribution both along the length and through the mesh,
(b) increase the value of the teeth by changing their form,
(c) increase the number of loads on the bull wheel.

Let us examine these possibilities in turn.

(a) We have seen how to achieve better load distribution along the teeth in section 7.2, and reference 36 gives an impressive account of improvements effected in naval gears in this way. Design for the development of the full value by controlling the variation of the load as the teeth pass through the mesh is described in reference 20.

(b) Section 6.9 discusses the problem of increasing the value of involute spur gear teeth, and much of this will read across into helical gearing. We might also consider a generalized tooth form, not necessarily involute, and maximize its strengths, a procedure which leads inexorably to short-path and point-contact gears, of which the Wildhaber-Novikov forms are typical [29]. Such tooth forms have a large inherent advantage in strength which can be exploited in at least two ways.

We can either make the gears much smaller, or make them of much weaker and cheaper stuff, or we can follow a middle course of smaller gears of somewhat weaker stuff. Here it is important to note that the load capacity of a given tooth contact is roughly proportional to the square of the material strength (section 4.8), so that as we use cheaper stuff, surface strength is reduced more than bending strength, and it is the latter which is more difficult to provide in circular arc tooth forms.

(c) If the number of loads on the bull wheel can be increased, its value increases almost in proportion. If we provide two pinions driving the bull wheel, we can reduce the facewidth to about 55 per cent—the odd 5 per cent being needed because the required life in cycles is doubled. If only bending strength is concerned, we can win back 2 per cent by rebalancing strengths between wheel and pinion as explained in section 6.9, but this is not possible with surface strength. An allowance for imperfect balance of loads between the pinions should be included—this is very large in marine reduction gears (the writer suspects the allowance is excessive and a vestige of ancient prejudices, especially since the longitudinal distribution of load will normally be improved).

Just to reduce the facewidth would be to squander the enhanced value; in accordance with the principle of section 3.3, that any change must be followed through in all its consequences, the reduction should be shared between the facewidth, the diameters and probably the tooth pitch.

Notice that in such a change the facewidth/diameter ratio might be expected to be little altered, or perhaps slightly reduced. If it was optimum before, it should be optimum after, from scaling considerations (section 4.7), but since casing size will now have a reduced effect there may be a slight shift towards a relatively narrower gear. Thus of the 45 per cent reduction rather less than two-thirds may go on (diameter squared) and rather more than one-third on facewidth, giving perhaps 0.83 of the diameter with one mesh and 0.80 of the facewidth ($0.83^2 \times 0.80 \simeq 0.55$). Some advantages may be expected due to improvement of the longitudinal distribution of load.

Using twin layshafts, multiple layshafts, epicyclics or star gears will reduce the gross amount of gear, but it will add to the prime cost due to all the extra components involved. Accordingly multiple meshing will only pay under certain favourable conditions, such as:

(a) very large tasks (output torques),
(b) a premium on low weight or size (aircraft, mining machinery, rail and road transmissions),
(c) large-scale production, which often favours the use of numerous smaller components.

A point which should be made about epicyclic gears is that the value of the annulus gear (its task-performing capacity or torque capacity) is less than the output torque T_0 by the amount of the input torque. This is not the case with star gears, which do not, however, suffer from centrifugal loading of the planet bearing, which is severe in such designs as that shown in Figure 7.7; in that particular case, centrifugal loads add to the stringency of the design problem set by the elastic diaphragm.

An extreme case of multiple meshing was the simple epicyclic reduction gear of the Wright 'Turbo-Cyclone' turbo-supercharged piston aeroengine, with 18 planet pinions.

8.3 Gears for contra-rotating propellers

It was mentioned in section 5.4 that contra-rotating propellers for ships showed advantages in propulsive efficiency. Let us consider the likely cost of gears for such propellers in the light of the last section.

It will be assumed that the propeller speeds are the same as for the original single screw, and that the shaft power is unaltered. Then if T_0 was the output torque for the original gearing, the new gearing puts out equal and opposite torques of $\frac{1}{2}T_0$. Are we to assess the task at $\frac{1}{2}T_0$, or $2 \times \frac{1}{2}T_0$?

The answer to this question depends on how we generate our two 'half torques.' If it is done in separate bull wheels separately driven, the answer is $2 \times \frac{1}{2}T_0$, but if we make one 'half torque' the reaction to the other, the answer is $\frac{1}{2}T_0$.

We have only two viable means of producing these torques in the propeller shafts—by loads on gear teeth attached to them, or by loads on bearings of planetary shafts attached to them. In the simple epicyclic or star gear we have both—the annulus-bull wheel, and a set of orbiting or potentially orbiting shafts. One of these torque capacities—the 'spider' of the star gear or the annulus of the epicyclic—has a large untapped value, an asset we can now

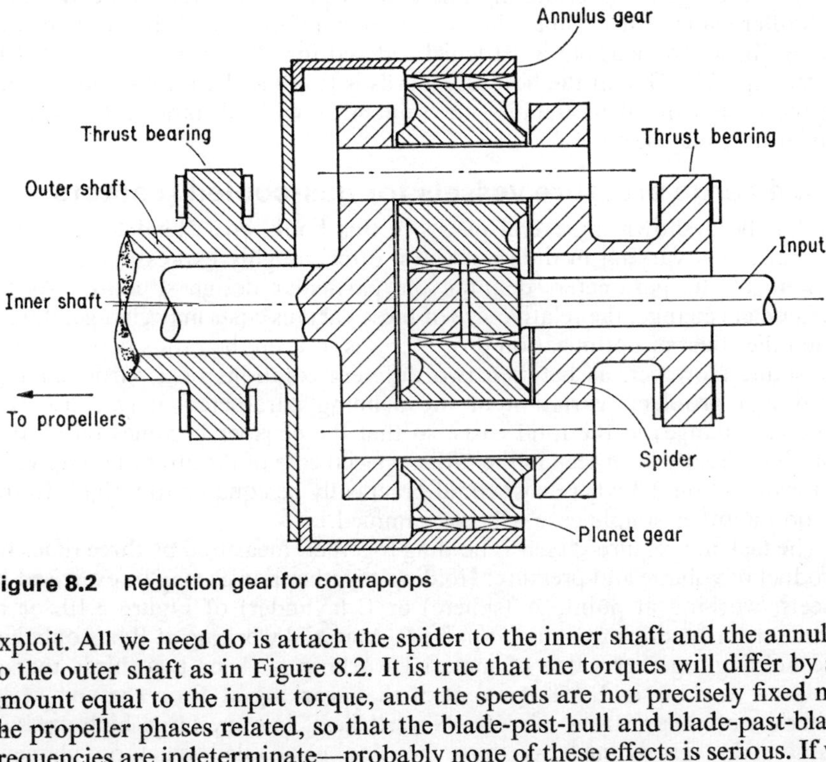

Figure 8.2 Reduction gear for contraprops

exploit. All we need do is attach the spider to the inner shaft and the annulus to the outer shaft as in Figure 8.2. It is true that the torques will differ by an amount equal to the input torque, and the speeds are not precisely fixed nor the propeller phases related, so that the blade-past-hull and blade-past-blade frequencies are indeterminate—probably none of these effects is serious. If we were worried about any of them they could be rectified, at a cost.

The task of the contraprop gearing in an arrangement like Figure 8.2 is only about $\frac{1}{2}T_0$, so that on straight gear cost we should expect a substantial decrease. The costs of shafts and thrust bearings will be higher than for the single screw, and some of the difficulties of assembly etc. will be reflected in the casing and spider costs. Above all, the structural-cum-topological difficulties of the star gear are bound to be expensive; in particular, one of these difficulties is easily overlooked, and it is worth digressing a little to examine it.

It is practically essential to have some bearing between the inner and outer shafts. Unfortunately, plain bearings between contra-rotating shafts do not work when, as in this case, the principal load does not rotate (here it is the deadweight of the shaft and propeller). This may be seen by giving the whole system a rotation so as to make the outer shaft stationary. The load is then rotating at half the speed of the inner shaft, and will force a half-speed whirl. (If we take the Reynold's equation as a starting point, it is readily seen that neither squeeze nor sliding wedge action is present.) One way round this problem is an intermediate stationary shell, but with the arrangement of Figure 8.2 it is awkward to prevent it rotating.

To guard against overlooking this kind of problem is very difficult; check lists offer some protection against the more common errors, like designs that cannot be assembled, but a list which catered for all contingencies would be unmanageable. One of the best safeguards is to try and visualize the working of the design in all its details—to 'think through' all its aspects, however trivial they may seem.

8.4 Large pressure vessels for gas-cooled reactors

As has been shown in several places in this book, costs tend to consist of several terms varying in different ways, often roughly expressible as simple powers of the parameters defining the particular design situation. As the parameters change, the relative size of these various costs may change, though often the effect of optimization procedures is to keep the ratios between them constant. However, as between two different solutions of a particular type of design problem, variation of the defining parameters may cause very different changes in the total costs, so that the cheaper becomes the dearer. This has already been seen in the rather special case of the aircraft compressor rotor in section 2.3 where cost could practically be equated to weight. In this section another simple case will be examined.

The task in pressure vessels is holding together, measured by three times the product of volume and pressure. Holding-together power may be exercised by sheets, working at points A (sphere) or E (cylinder) of Figure 3.10, or by filaments or tendons working at point B. Material working at B has only half the holding-together value of material working at A or two-thirds that of material working at E (58 per cent of that working at E on the von Mises criterion of shear strain energy). Nevertheless, in very large high-pressure (large-task) vessels such as those housing gas-cooled reactors, filament-strengthened vessels are cheaper than steel shells. So much cheaper are they that a different form of design with all the coolant circuit within one pressure vessel becomes viable, eliminating the difficult thermal-expansion problems mentioned in section 7.4. (This solution also involves an example of separation of function, referred to in section 2.12.) A contributory factor in the superiority of the prestressed reinforced-concrete vessel is the high value/cost ratio (for material at point B) possible in steel-wire reinforcing tendons. But above all, the sheer size of the task favours the tendon-strengthened vessel in the following way.

The amount of material required for either type of vessel is roughly proportional to the task, or to pL^3 where p is the gauge pressure and L a typical dimension, so that a crude algebraic expression for the cost should contain such a term. However, fabrication and erection costs exceed material costs and here an important factor is the thickness of the plates of which the steel shell must be made, which varies as pL. Thick plates become very expensive to weld, so that we can expect the cost per ton of the steel-work to rise fairly fast with pL. On the other hand, the placing, anchorage and prestressing costs of each tendon will not increase much with its length, so that the total for all the tendons will increase little faster than pL^2.

It is to be expected, therefore, that the cost per unit of holding-together power of the welded steel shell will rise with the task, while that of the reinforced-concrete vessel will fall. For the same reason, it is cheaper to subdivide a steel shell into a number of smaller vessels connected by ducts, the minimum size being dictated by the largest indivisible unit, the reactor itself, while the most economical arrangement with reinforced concrete is a single all-enclosing vessel.

Based on the designed ultimate pressure of 1155 p.s.i.g. (7.9 N mm^{-2}), the task of the Oldbury 'B' Vessel, of about 279 000 ft^3 (7900 m^3), is 1.67×10^{12} lbf in. (189 GJ). At the ultimate load of the anchorages the holding-together power of the tendons is about 4.1×10^{12} lbf in. (460 GJ). The discrepancy is chiefly due to the thickness of the walls, which adds to the effective task; the median surface of the walls encloses about twice the volume of the vessel itself (this ratio would be less in a more modern design).

9 Various Principles and Approaches

9.1 Introduction

This chapter collects together various helpful ideas. They are of fairly general application but too vague to merit the title 'methods', and they are not dealt with at length.

9.2 Avoiding arbitrary decisions: degrees of freedom of choice

Every choice with which the designer is faced represents an opportunity, a *degree of freedom* in the design which can be exercised to improve it. An arbitrary decision represents a wasted opportunity. A. E. Moulton has explained in a lecture how, when he designed his famous small-wheeled bicycle, he examined each of the decisions implicit in the design of the 'traditional' bicycle and considered the practicability of alternatives at each stage.

There are two difficulties in avoiding arbitrary decisions; the first is to recognize when a decision is being made. That this is not always easy is shown by the following example.

In the early days of the aircraft gas turbine, it was impracticable to make the turbine disc and blades in one piece. (It still is, in most applications.) The practical way was to make each blade a separate piece and fix a set of blades to a disc; the preferred form of fixing was the fir-tree root. All this was sound enough. The arbitrary decision which passed unnoticed was that the fixing was put at a radius just slightly less than that of the inner end of the aerofoil, as in Figure 9.1(a). The result of this was that the disc rim was at a temperature little less than that of the blade itself, restricting the choice of disc materials to those having refractory properties nearly as good as those of the 'Nimonic' alloys used for the blades. This meant the use of austenitic steels which were expensive and dense, and weak and unreliable compared with the ferritic steels which might have been preferred were it not for the high rim temperature. A further disadvantage of the austenitic steel was a high coefficient of thermal expansion.

A great improvement was made by Rolls-Royce, who put the fixing at a smaller radius. The truncated wedge of blade material left between the aerofoil and the fixing, sometimes called an extended root, was hollowed out on either side to leave an I section (Figure 9.1(b)). By this means the maximum temperature in the disc was reduced to a point at which it was possible to use creep-resisting ferritic steels. The combined and roughly equal effects of lower temperature difference between hub and rim and lower coefficient of thermal expansion reduced the thermal stresses to about one-third in some cases (section 4.2), and this and the higher strength of the ferritic steel increased the effective holding-together power (section 3.12) more than twice.

By this change, reductions in the weight of bladed discs of up to 55 per cent were effected. The fir-trees had to sustain a higher load on a smaller pitch

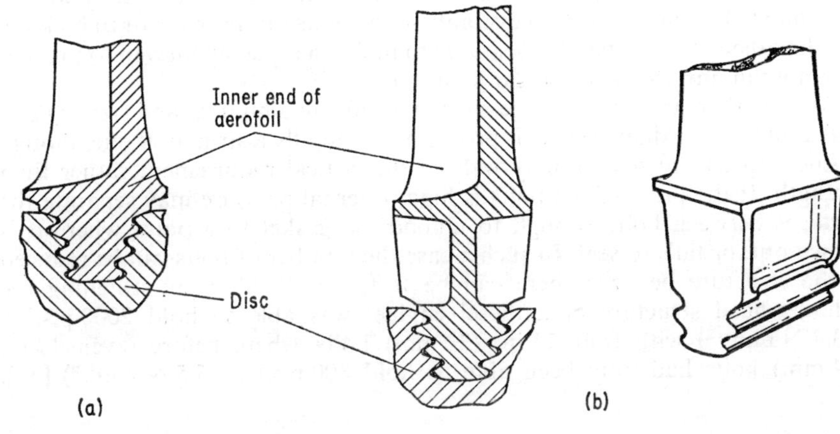

Figure 9.1 Extended root turbine blade

length, but this was made possible by the better properties of the material of the female part. Savings on the cost of the disc forgings were offset by the greater length of stock required for the blades, and slightly more cooling air may have been needed, but the saving on total engine weight was a great improvement.

While it is easy to be wise after the event, the delay of almost a decade in arriving at this solution could scarcely have occurred had a sufficiently abstract view of the problem been taken in the first case, especially if this particular variable, the radius of the fixing, had been recognized as such. Notice the role of established ideas, the blade and the disc, and the example of steam-turbine practice, in obscuring the important freedom of choice.

Once the impracticability of making the wheel (bladed disc) in one piece is established, and the likelihood that it will be profitable to use different materials for different parts is recognized, then the abstract view of the problem is that of making one part (a wheel) in two materials A and B. It is then automatic to regard not only the natures of A and B but also the position of the boundary between them as a matter for choice. It is for this sort of reason that it is advocated that the designer should continually increase the level of abstraction at which he looks at his problems, at least to the point where he is tolerably sure that the process has become unprofitable.

The second difficulty is more conspicuous in the embodiment than the conceptual stage; it is that of deciding how to take advantage of the degree of freedom of choice once it has been recognized. Sometimes it is difficult to see any way in which it can be useful, and sometimes it is difficult to be sure that its greatest potential use has not been overlooked.

As a rather trivial example, consider the location of six bolt holes on a given pitch circle in a flanged joint. They may be made equally spaced, and this is the best arrangement from a structural point of view. This still leaves a degree of freedom in the angular position of the set of holes, a detail which may well be determined by studying the accessibility of the bolts for insertion or tightening. On the other hand, it may be desirable to make the spacing uneven to prevent assembly in an incorrect angular position.

At an earlier stage, it was necessary to decide on a number and size of bolts and a pitch circle diameter. This problem is generally a structural one, though stiffness to ensure sealing may well be the critical requirement rather than strength. If the pitch of the bolts is high, internal pressure may separate the surfaces between bolts enough to unload the gasket to a point where it is blown out, or fails to seal. In such a case the number of bolts can be reduced if the structure between them can be stiffened. Feilden, by stiffening the cylinder-head structure of a diesel engine, was able to hold 2000 p.s.i.g. (13.8 N mm^{-2}) with four $1\frac{1}{4}$ in. (32 mm) bolts where before seven $1\frac{3}{4}$ in. (44 mm) bolts had only been able to hold 800 p.s.i.g. (5·5 N mm^{-2}) [37].

9.3 Mathematical models

It might be thought that the mathematical model is too familiar a notion to require special mention in a book of this sort. Nevertheless, designers are apt to restrict their use of mathematical models to those they have met in text-books or technical papers. In the cylinder-head design problem just discussed, the failure to use a simple mathematical model was a root cause of the original weakness. The good designer will try to devise mathematical models of his own, even in the least promising situations. As an extreme example it is perhaps fair to mention Lanchester's pioneer work on the operational theory of warfare.

To illustrate the application of mathematical models to an unlikely field, the design of beds will be considered—beds for human beings, not machines. Market research may return rather subjective statements of preference—'comfortable but free from sagging' or 'soft yet firm'. How can we find useful measures or concepts (section 4.5) that may be related to these qualities?

For the bed we substitute a mathematical model in the form of an elastic foundation. To simplify the treatment, it will be restricted to a thin vertical slice of bed taken parallel to its length, so the problem is in two dimensions only, length and depth. An axis of x is taken in the unloaded level surface of the bed, and an axis of y downwards (Figure 9.2). In the unloaded condition the surface of the bed is given by $y = 0$, and under a loading w per unit length, where w is an arbitrary function of x, it is given by

$$y = \mathrm{F}(w, x)$$

It is more convenient for our purpose to invert this relationship to give w in terms of y,

$$w = \mathrm{f}(x, y) \tag{9.1}$$

Now this function f describes the stiffness of the bed surface, and it is the designer's task to ensure that it has as nearly as possible the most suitable

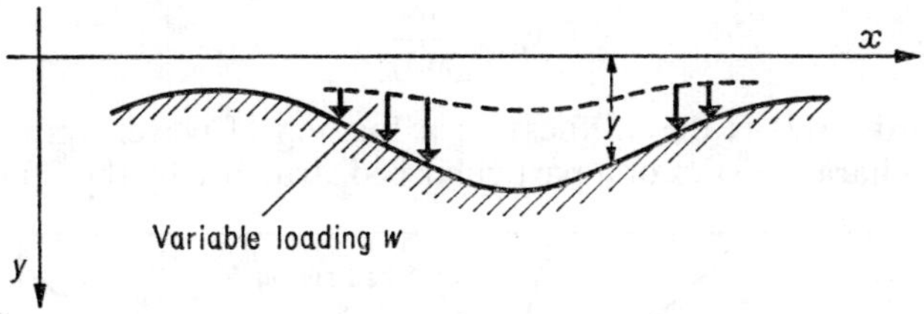

Figure 9.2 Loads on idealized bed

form, whatever that may be. Let us see what can be done with the excessively general and unintelligible equation 9.1.

Firstly, let us assume that the bed is uniform throughout its length and that the ends do not greatly differ in their properties from the rest. Then a shift of the origin of x can make no difference in the relationship of w and y, so that x cannot occur explicitly in equation 9.1, which may be rewritten

$$w = \mathrm{f}\left(y, \frac{\mathrm{d}y}{\mathrm{d}x}, \frac{\mathrm{d}^2y}{\mathrm{d}x^2} \cdots\right) \tag{9.2}$$

Secondly, let us assume that the bed is linear; then equation 9.1 can be rewritten

$$w = k_0 y - k_2 \frac{\mathrm{d}^2y}{\mathrm{d}x^2} + k_4 \frac{\mathrm{d}^4y}{\mathrm{d}x^4} - \ldots \tag{9.3}$$

Odd-order derivatives cannot occur because their sign would reverse if the direction of x was reversed. If

$$y = y_0 \cos\left(\frac{2\pi}{\lambda} x + \phi\right)$$

substitution in equation 9.3 yields

$$w = y_0\left[k_0 + \left(\frac{2\pi}{\lambda}\right)^2 k_2 + \left(\frac{2\pi}{\lambda}\right)^4 k_4 + \ldots\right]\cos\left(\frac{2\pi}{\lambda} x + \phi\right)$$

$$= y_0 s(\lambda) \cos\left(\frac{2\pi}{\lambda} x + \phi\right) \tag{9.4}$$

where $s(\lambda)$ is the expression in square brackets. The right-hand side of equation 9.4 can also be written in the form

$$w_0 \cos\left(\frac{2\pi}{\lambda} x + \phi\right)$$

where

$$w_0 = y_0 s(\lambda) \tag{9.5}$$

Then equation 9.4 can be interpreted as follows: a sinusoidal load of amplitude w_0 and wavelength λ produces a sinusoidal deflection of the same phase and wavelength and of amplitude

$$y_0 = \frac{w_0}{s(\lambda)}$$

In other words, $s(\lambda)$ is the stiffness to a loading of wavelength λ. We can describe the characteristics of a particular bed structure by the function s. If s

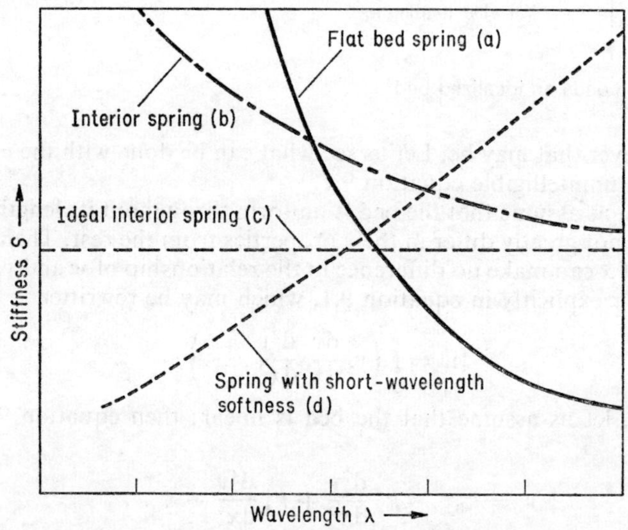

Figure 9.3 Stiffness wavelength relations for beds (hypothetical)

increases with λ, the bed is stiffer to long-wavelength loadings than to short-wavelength ones.

Now let us think what this means in terms of comfort. To support the sleeper with his spine free from flexure, a relatively high stiffness is required at wavelengths of about 3 to 6 ft. At wavelengths of 1 ft or less a lower stiffness is desirable so that local bumps on the sleeper, like arms, shoulders and haunches, can sink easily into the bed without developing high local pressures; many people will have felt at some time they would like a hole in the bed to slip their arm into, if only that hole could always be in the right place.

Flat bed springs, hammocks and such approximate to the case where only k_2 in equation 9.3 has a significant value, so that s is proportional to λ^{-2} (see Figure 9.3).

The separate coil springs of the conventional interior spring matress ideally have all the k's in equation 9.3 zero except k_0, so that s is constant. The covers of such mattresses and 'spring bases' are generally sufficiently inextensible to introduce a measure of k_2 however, so the practical curve of s against λ would probably show a considerable downward trend, while what is needed is an upward one—'short-wavelength softness' might be our slogan. It seems to the writer that some designs of continuous spring may possess in themselves a measure of short-wavelength softness, but this is probably swamped by cover tensions.

One final point is that the mathematical model enables us to change our viewpoint (section 4.10). Having decided on the curve of s against λ we want (perhaps as in Figure 9.3) we can derive the deflection under a point or very local load, as in Figure 9.4. Notice that the 'short-wavelength softness' bed

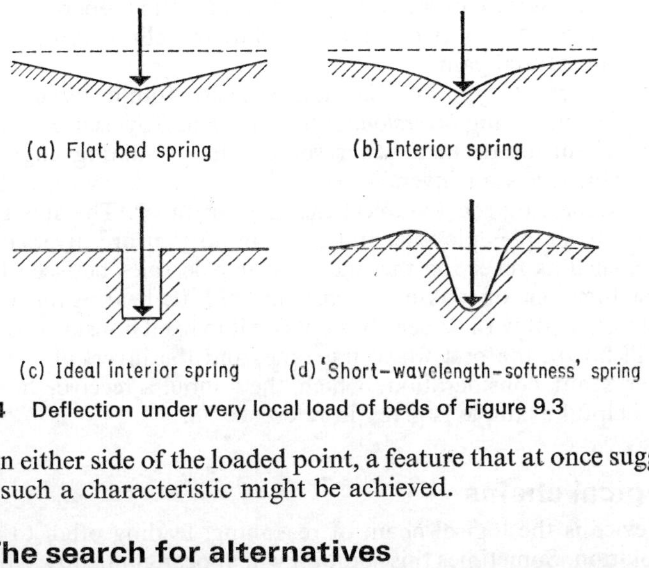

Figure 9.4 Deflection under very local load of beds of Figure 9.3

rises up on either side of the loaded point, a feature that at once suggests ways in which such a characteristic might be achieved.

9.4 The search for alternatives

The importance of looking for alternative ways of doing things is well recognized and has received attention elsewhere in this book. Even in a combinative treatment, however, possibilities may be overlooked. A good comprehensive list of known engineering means would help, a standard work on the engineering repertoire (section 2.3).

Books of various kinds exist, indeed, but they are generally limited in scope, more detailed than is necessary for the present purpose and not abstract enough in their classification; they tend to overlook the kind of alternative the designer overlooks, through their concern with types of device, rather than purposes.

As a simple example, suppose we look through a list of materials to find one of high thermal conductivity; it is unlikely we shall find quoted the extremely high effective values given by various forms of 'natural' convection—a single phase of sodium or two phases of water in a high centrifugal field, the steam tube in a baker's oven, a 'heat pipe' where capillary action returns the liquid phase or perhaps a magnetic fluid in a suitable magnetic field. Yet one of these might suit your purpose.

One particular alternative that can always be considered and occasionally leads to a profitable idea is to dispense with a function altogether. As an example, a small team working under the writer's guidance on the design of a very small heat engine using the Rankine cycle found that one of the difficulties was the feed pump. They finally put the condenser above the boiler and arranged a float valve to lift and put the condenser in direct communication with the boiler when most of the working fluid had accumulated as liquid below the condenser. This liquid ran down into the boiler, now depressurized, the valve was closed, and the turbine started again when the boiler boiled once more. The extra irreversibilities and intermittent operation were no great disadvantage in this particular case, but the elimination of the feed pump was a substantial gain.

Another important type of alternative easily found by a systematic approach is that involving *inversion*, either in the narrow sense as applied to academic mechanisms or in a wider sense. The oscillating engine used a mechanism which was an inversion of the slider-crank chain, obtained by making the connecting rod (coupler) the fixed member. The star gear is an inversion of the simple epicyclic. But more important are inversions in the wider sense, such as reversing the male and female roles between two parts which screw together, or making a shaft run with its bearings inside it on a central pin, instead of with its bearings outside it in fixed housings. Usually the designer will hit on the best form first time, and the inversions can be discarded after scant consideration, which they should receive, however. A simple but helpful example is given in reference 38.

9.5 Logical chains

A useful device is the logical chain of reasoning, leading often to a unique practical solution. Sometimes this solution will appear eminently satisfactory, and sometimes it will seem unduly clumsy or expensive. In either event it is worth testing the chain link by link in the hope of finding some fault in the reasoning; the good solution may be bettered, and the poor solution proved not to be unique.

As an example of an idea originating in a logical chain, the writer was asked some years ago what use could be made of the cold sink available when liquid natural gas shipped in from Algeria was evaporated and warmed to ambient temperature preparatory to use. Various suggestions had been made that it be used as the cold sink of a heat engine. The first step in gaining insight is to recognize that the potential money value of one joule of available energy of a liquid gas as energy is less than that of one joule of available energy in a common fuel, because of the special, and hence expensive, plant needed to convert it. The money cost of producing the available energy (or cold, for short) is relatively high because of the special, and hence expensive, plant required to liquefy the gas. It follows that if a really profitable use is to be found it must be as *cold*, not simply as *available energy*.

The next step is to enquire what existing uses there are of this quality of cold in such large quantities, just as might be asked of any material commodity. The simple answer is easily missed; the same quality and quantity is used in liquefying the gas for shipment. If the cold could be shipped back, it could be used to liquefy more natural gas.

The only way to ship the cold back is in some substance; the most attractive possibility is a liquid gas, but the choice is limited since it must either be a genuine return cargo or something of very low cost. The only gases which are cheap enough are those present in large proportions in the air, and safety considerations rule out oxygen. We are left with nitrogen as the only possibility.

Let us see where we have arrived. A ship loads L.N.G. at A say, and sails to B where it unloads L.N.G. and takes on liquid nitrogen. During the interval before the arrival of the next ship a conversion plant at B vaporizes and warms L.N.G. and separates and liquefies atmospheric nitrogen—the separated oxygen gas can perhaps be sold.

By the time the next ship arrives, there is a fresh cargo of liquid nitrogen ready. At A there is a plant which carries out the reverse process, liquefying natural gas using the cold in the liquid nitrogen, which emerges at atmospheric temperature and pressure.

Because of evaporation en route and irreversibilities in the plants, the process cannot be self-sustaining, so that at one or both ends there must be a substantial power input. The amount of power needed depends on the difficulty of matching nitrogen to methane, a problem rather like that of the dual-pressure steam cycle in section 5.7. It transpires that the difficulty is great, and this together with the loss of valuable flexibility in the rate of vaporization of L.N.G. at B make the idea less attractive. Nevertheless, it was worth studying.

The most important step here was the recognition of the potential use of the cold arising from giving the right answer to the question: 'Who uses this quantity and quality of cold?'. A question and answer approach is often helpful, and valuable questions often seem childish at first. Another important step was recognition that the cold could only be of much value as cold.

As another example, consider the problem of the slow heating up of the rings on electric cookers. The rings are resistance heating elements that have to be robust since they are in contact with the pans and subject to fairly rough usage. They have also to be electrically safe, though exposed, and to provide a sufficient surface area for heat transfer to the pans. These requirements impose a tight lower bound on the volume and hence the water equivalent of the rings, and this in turn means a long heating-up time.

This logical chain has at least three weak links. Firstly, it is not essential to use resistance heating outside the pan—we could use microwaves, for example. Secondly, it is not necessary to have the rings in contact with the pans—we could use an intermediate heat transfer agency. Thirdly, the long heating-up time is not absolutely necessary given a high water equivalent, since we might be able to boost the power to begin with. None of the three seems very promising but a combination of the second and third is attractive.

9.6 Past practice and changed circumstances

Every designer automatically and necessarily draws very largely on what has been done before. It is difficult to do this without accepting as essential things which are not, making provision for circumstances which do not arise, or adopting practices which are no longer, or never were, the best available. Some defence against these sorts of error is provided by a systematic review of the logic behind the scheme in question and a diligent search for viable alternatives, as briefly discussed in the last few sections.

A designer tackling a problem outside his own field often has a freshness of approach and a freedom from preconceived ideas that is of great value. It has been argued that this aspect is so important he should deliberately avoid acquainting himself with past practice, but start absolutely from first principles as if he were the first man to work in the field in question. The writer feels this is too extreme an attitude, and that it is possible to retain the advantages of starting from scratch while securing the valuable insight which can often be gained from a study of past practice. Since gifted designers are often of great individuality, not to say perversity, it is not likely that they will be blinded to alternatives by orthodox practices without a long period of habituation.

9.7 Brainstorming

A useful technique is that called *brainstorming* [39], in which a group of designers sit down to produce as long a list of 'solutions' to the problem in hand as they can in a given time. The essence of the method is that no analysis of the 'solutions' is attempted at the time, though the writer's limited but encouraging experience suggests that this rule should not be strictly adhered to.

Each designer (or other brainstormer) contributes a tentative idea as soon as it comes into his head, and this will often stimulate a chain of further suggestions from his colleagues, generating a kind of momentum which can easily be lost if too much time is given to any particular solution. The end product of the 'session' is a host of tentative ideas, some of which may be valuable.

It is usually said that no criticism should be allowed, either because it is inhibiting or because it breaks the flow of ideas. The writer, however, agrees with Dixon [40] that short objections to suggestions often prompt the next one, and in any case, the members of the group need to be got into a frame of mind where they are not inhibited by criticism.

A tape-recorder is useful—note-taking slows the process up for at least one member. However, it is useful for the chairman to make notes of any point at which there was a tendency for the development of ideas to branch. Then, when the chosen branch peters out, he can bring the session back to the alternative line. In a thriving brainstorm there are always too many directions to follow all at once, and a 'reprise' will often restore the momentum. Above all, no-one should be allowed more than, say, three sentences at a time (unless his idea is particularly complicated).

Reference [40] is excellent on this topic.

Questions

Q.9.1(2). Figure 9.5 shows the geared elements of an automatic transmission. Gears 7 and 8 are welded together and are free to turn on the pin through

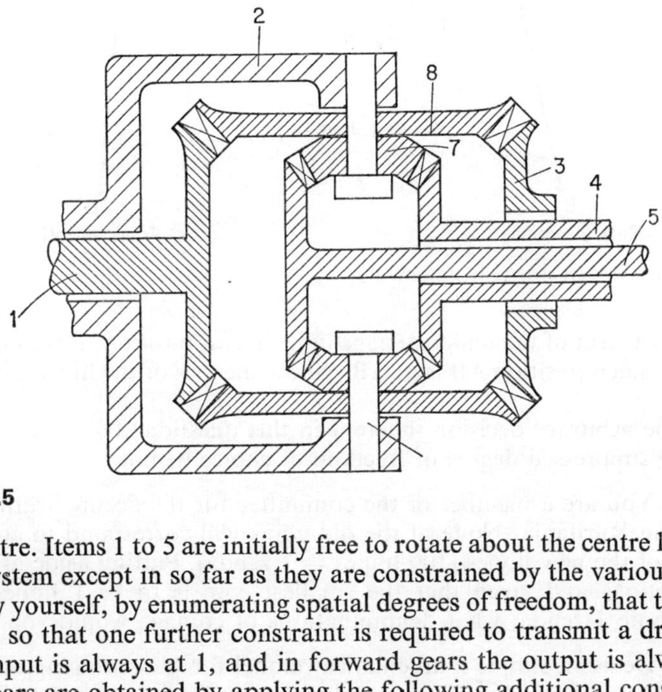

Figure 9.5

their centre. Items 1 to 5 are initially free to rotate about the centre line of the whole system except in so far as they are constrained by the various meshes.

Satisfy yourself, by enumerating spatial degrees of freedom, that the system has two, so that one further constraint is required to transmit a drive.

The input is always at 1, and in forward gears the output is always at 5. These gears are obtained by applying the following additional constraints.

First, hold 2	Third, hold 3
Second, hold 4	Top, lock 4 and 5 together

Show that there is only one degree of freedom of choice in fixing the set of gear ratios, i.e. that if any gear ratio (except top, which is 1:1) is fixed, all the others are fixed.

Find the concealed arbitrary decision in this design; i.e. find the change in geometry that can be made without altering the topological or essential nature of the gearing system that will increase the number of degrees of freedom to two, so that two gears (other than top) can be freely chosen (within the limits imposed by practical numbers of teeth). (This is the basic gear system of the Automotive Products automatic transmission used by B.M.C. The concealed degree of freedom was observed by the designers, who decided not to use it for reasons given in reference 41: although superficially these reasons are excellent, the writer believes they conceal another arbitrary decision, that of Question 9.5, in effect.)

Q.9.2(3). Figure 9.6 shows two views of an aircraft undercarriage used to illustrate a problem in descriptive geometry, which was as follows: AB

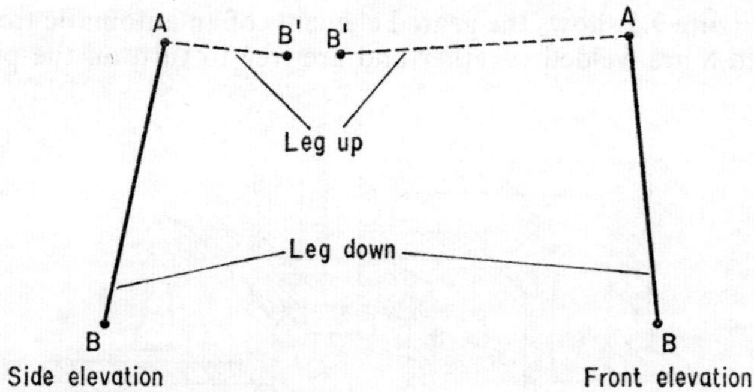

Figure 9.6 Undercarriage leg geometry

represents a strut of the undercarriage in the down position: it is hinged at A. In the retracted position AB is at AB′. Show the axis of the hinge pin in both views.

Find the arbitrary decision inherent in this question. To what sort of use might the suppressed degree of freedom of choice be put?

Q.9.3(2). You are a member of the committee for the decimalization of the currency in Ruritania. None of the old coins will correspond to convenient numbers of the new hopes (100 hopes = 1 zenda). Putting aside all preconceived notions and given that the smallest coin is to be 1 hope and the smallest note 1 zenda, what denominations of coinage would you suggest?

Q.9.4(3). Devise a simple mathematical model for the springing of the rear wheels of a car which will fit solid axles, independent suspension, and independent suspension with an anti-roll bar.

Q.9.5(3). It is sometimes stated in works on gearing that the number of teeth in the planet gears of a simple epicyclic gear must be half the difference of the number of teeth in the annulus gear and the sun pinion. This is a fallacy. Why?

Answers

A.9.1. The axis of the gear cluster 7 + 8 need not be perpendicular to the axis of the system.

A.9.2. It is arbitrarily decided that the hinge axis must be perpendicular to the strut. In fact, any hinge line through A and lying in the plane bisecting BB′ at right-angles will do, i.e. there is a degree of freedom of choice which might be used, for example, to make the wheel tuck flat into the wing.

A.9.3. It is an arbitrary decision that the denominations must be factors of 100. The most convenient choice of three denominations between 1 and 100 is 3, 10, and 30, requiring on average only about 80 per cent as many coins giving change and smaller numbers of one denomination to count, compared with 2, 5, and 20.

A.9.4. Turn the car upside down and apply a downward force F to one wheel. Suppose that that wheel descends a and the other one b. Then a/F is the *direct flexibility* and b/F the *cross-flexibility* and these two together define the properties of the suspension (in a simple treatment). The cross-flexibility is negative with a solid axle sprung inboard of the wheels, zero with an independent suspension and positive with an anti-roll bar on an independent suspension.

A.9.5. The fallacy arises from the arbitrary assumption that the pitch circles of the planet gear in its meshing with sun pinion and annulus must be the same circle. There is no reason at all why this should be the case—indeed, it is a special virtue of involute gear that the pitch circle is not fixed. The gears in Figure 7.7 have numbers of teeth 24, 47, and 120.

10
Conclusion and Select Bibliography

10.1 Resumé

The preceding chapters have dealt with various methods and approaches that may be helpful in conceptual design problems. Some of these, like the combinative studies of Chapter 2, are of general application. Others, like the identification and treatment of generic types of problems such as matching (Chapter 5) and disposition (Chapter 6), are only of use in special cases; nevertheless, as the list of topics in Chapter 5 will show, the special cases are so frequent that the approach is of great value. The insight-developing techniques of Chapter 4 are strongly recommended, and the writer believes that the investment of time in such exercises will be amply repaid in better designs more expeditiously arrived at. The advantages of rough calculation of orders of magnitude and 'quick sums' in particular cannot be over-emphasized; the optimizations dealt with in Chapter 3 are of this sketchy sort. The space available would not allow of fuller treatments, but in any case these simpler exercises yield perhaps nine-tenths of the possible benefits at a few per cent of the cost. Of course, the heavy calculations that fetch in the remaining tenth are still very worthwhile whenever the sums of money to be spent are large.

Chapter 7 (kinematic and elastic design) differs from the rest in the restricted physical nature of its subject matter, which nevertheless is of such importance as to warrant its inclusion (the writer abandoned a similar chapter on structures when it exceeded a third of the length of the present book). There are very few engineering artefacts of any size or importance which nowhere in their design warrant a consideration of the conflicting merits of kinematic and elastic alternatives.

Chapter 8 (tasks, costs, assets, and values) is inadequately developed and has a poor range of examples, but the writer confidently expects that in a decade or so this will be a well-established field of knowledge and a prized support of design and management of many kinds. The ideas of Chapter 9 will be familiar to many readers, but it is hoped they will find the examples interesting and valuable, and the whole a useful reminder.

Finally, the select bibliography which concludes this chapter suggests material for further study in all directions. It shows the catholic and liberal nature of engineering design which make it, properly pursued, intellectually satisfying and educational in the most demanding sense.

10.2 Checking and evaluation

Every designer is only too familiar with the pitfalls that beset him—an exceptionally treacherous one is described in section 8.3. The best protection against them is a complete review of all the processes of manufacture, assembly, operation, and maintenance of the product, and most of us have imaginations insufficiently vivid and complete not to benefit from the aid of a prompting check-list; here is one.

Function

Does it work?
Does it meet the specification?
Will threads seize owing to heat, rotors seize owing to transient thermal expansions, bearings overload for the same reason, bearings fail before oil reaches them or before they generate a film, will bearings brinell, will the starting torque be excessive, will thermal distortions or loads misalign bearings, will coolants freeze, will air, water, oil, dust, explosive mixtures, swarf, etc., collect in unsuitable places, will seals or sumps produce excessive drag, parts fret, nuts slacken, threads strip after repeated tightening, lubrication or fuel systems fail to function in certain positions, unequal tightening of bolts cause distortion, balance be lost due to stripping for assembly, bits fall into inaccessible places? (they will!)
Can it be got into its required location?

Manufacture

Can it be made?
Are the materials available, can they be treated and machined as required?
Can it be assembled? in only one way? without the help of periscopic-eyed dwarfs with socket-headed fingers?
Can the parts and the assembly be handled and transported? Are lifting eyes, lashing points, etc., required?
Are special tools needed for assembly or maintenance?
Are there alignment problems, requiring special equipment?
Are castings, forgings, welds, etc., within the capacity of the shop?
Is a practical sequence of heat treatment and machining possible?
Is any machining after assembly needed? and if so, is it possible?

Maintenance

Have inspection covers been provided?
Can bearing clearances, etc., be checked?
Have suitable instrumentation, oiling and greasing provisions been made?
Can parts likely to need replacement be replaced *in situ*?

Can worn parts be salvaged, if necessary with the replacement of other parts with oversize ones?
Can flushing and cleaning be done as required?

Strength

Have all loading conditions been considered?
Have all modes of failure been considered, i.e. overload, fatigue, creep, brittle fracture, fretting corrosion?
Have vibration, thermal stresses, impact loads been considered?
Have effects reducing strength been considered, i.e. temperature, tolerances, finish, chemical nature of environment, corrosion, variation in material properties, internal stresses, welding, stress raisers, wear, effects of fitting or maintenance?
Are differential expansions allowed for? transient as well as steady?
Could mistakes in operation in service have dangerous results?

Safety

Is it safe?
Have any safety regulations been complied with?
Are warning notices required? interlocks? warning devices? fire-extinguishing devices? overload devices? overspeed devices?
Is it electrically safe?
Has drainage provision been made?

The process of evaluation has been covered fairly well in discussing design itself, since many of the methods involve evaluation in their execution. Nevertheless, no mention has been made of the financial calculations needed to carry out many of these evaluations properly, and indeed this is too large a topic for this small book. References 42 and 43 are suitable reading for the designer who is concerned with products where the proper techniques make a substantial difference—that is to say, almost anything where the total value of the product to be made is sizeable and the life of the plant or production line is longer than about a year.

Finally, an important aspect of evaluation is the cost of the designing—it is not worth expending great ingenuity to reduce the cost of a one-off product by £100 or so. On the other hand, the saving of one new penny on a car door lock of which perhaps twenty million will be made will justify the employment of four good designers, each earning £4000 per annum, for five months on studies in parallel and series (the rough calculation allows for overheads, associated detailing and costing work, risks and a usurious rate of return). The writer suspects that the proportion of the cost of most products that goes into design is much too small, but of course a limit is set by the amount of developed talent available. This is why the industrial designer is invading the field of engineering design proper. He has brains, creativity, and a sketchy knowledge of engineering, and he is rushing in to fill the vacuum left by the lack of suitable engineering graduates.

10.3 Bibliography: general works on design

Matousek's book [44] is in many ways complementary to this one, dealing chiefly with detail design (though in several places it touches on conceptual design). It has many pages of illustrations of detail, good and bad, right and wrong, and should be a great help in developing sound practices.

Rodenacker's *Systematic Design Method* [45] has been mentioned in section 2.6 as an example of a combinative approach, but it is much more than that. It comes a great deal nearer to a fully formalized approach than this book, and deserves the serious study of any designer who aims at excellence. In its combinative aspects it deals with both functions and configurations, but it has also an underlying methodology, which is an inversion of the scientific method. Its value has been amply demonstrated by use.

Marples's [46] treatment of design as decision making is a classic and well worth careful study. It has special importance with regard to the management of design and the case histories are of great interest. Marples's decision-tree approach is excellent for mapping the alternatives of the design process and steering a course amongst them.

The abundant literature on design is well recorded over a very fertile period in the *Design Abstracts*, compiled and published by Enfield College of Technology between October 1964 and January 1967, and since then by the Institution of Engineering Designers in its journal, *The Engineering Designer*.

A historic work, but one of continuing interest to the designer, is *Kinematics of Machinery* by F. Reuleaux [47]. His classification of mechanisms, though no longer used, contributed to Wankel's formulation of his search for chamber mechanisms suitable for use in i.c. engines, which culminated in the Wankel engine (section 2.6). Reuleaux's approach is a valuable example of how to come to grips with the sort of slippery problems designers face.

Finally, an excellent survey by Jones [48] is well worth reading for the lucid summaries it gives of methods extant in 1966.

10.4 Bibliography: special aspects of design

Torroja's works on structures must surely be among the most valuable and authoritative works on design. Particularly, the writer delights in the following passage on insight (Chapter 4) from '*Razon y ser de los tipos estructurales*' [49] which might be translated, 'the raison d'être of the various types of structure'.

> It is not enough to have studied structural theories and the calculations associated with them: it is necessary to have thought deeply and worked with them all to the extent of feeling as something right, natural and inbuilt the phenomena of stress and strain, so as to grasp as a whole how a structure works, and how it would fail, so that all this appears to you with the same clarity and conviction as that with which you apprehend the fall of a stone through space or the inexorable force which drives the arrow from the crossbow.

Torroja deplores in his opening paragraph the relatively few books on structural design, as against the many on structural theory, and goes on to show with simple but elegant examples, largely from his own work, the few but richly interacting and combining elements of the structural repertoire.

The Design of Structures of Least Weight, by H. L. Cox [17] is a difficult book to read, very concise in form (it is full of the kind of difficulty exemplified by Question 10.15). However, there is real value in it for those who can master its complexities. It starts by dealing with components, and this part is full of insights into structural economy. At the end it tackles the 'layout' or, as the writer would say, 'configuration', problem and gains a precarious grip on this elusive topic.

A valuable series of practical studies in the application of the sort of ideas discussed in Chapter 7 is given in reference 50.

10.5 Bibliography: case studies

The library of case studies built up by H. O. Fuchs and K. H. Vesper at Stanford University is a great resource, though intended primarily for students.

Feilden's paper [37] is of the highest value, full of typical cases and a counterblast to the 'blind-them-with-science brigade', those who cannot see the wood for the trees themselves and would have everyone else in the same sorry state.

While disagreeing profoundly with his conclusions, the writer recommends the paper by Booker [51] as a well-observed case study.

10.6 Bibliography: background reading

The writer believes that ecology, the study of the natural economy of living creatures, is of value to designers. An important concept in ecology is the *niche*—a gap in the pattern of dependence and competition of species of plants and animals that can accommodate another species. In just the same way, a product such as a motor car needs to have an *ecological niche* in the market. Car B may be similar in price and performance to an established car A and a little inferior in roominess, but it has a more sporting image and front-wheel drive, so it has its own kind of customer. Reading works on ecology will help to keep the mind of the designer open to a wide range of possibilities.

Besides ecology, the study of the design of living creatures and plants is full of lessons for the engineer. The adaptation of form to purpose is there seen in the most sophisticated guises. *Of Growth and Form* by D'Arcy Thompson is a remarkable work and one which is a delight to read, apart from its special value and interest to the engineer (the abridged form [52] is adequate for all but the enthusiast).

A rewarding study for designers is the engineering sketches of Leonardo da Vinci. Here the relationship between draughtsmanship and design reaches rare aesthetic heights, and the verve of the drawing comes from the kind of insight that Torroja declares the engineer must strive for. Something of the same surety and vigour, born perhaps of an even higher perfection of apprehension, is to be found in the sketches of the great engineers, e.g. I. K. Brunel.

Quite another kind of background is provided by Polya's book [53], which deals with problem solving in a general way, and is a source of many helpful ideas of the kind discussed in Chapter 9.

Questions

Q.10.1. A vertical take-off and landing (v.t.o.l.) aircraft has a wing which can be rotated so that its leading edge points upward. Four propellers ranged along this wing are driven by four gas turbines and interconnected by shafts so that all rotate at the same speed and continue to do so even if one turbine fails. Study possible arrangements of the engines, gearing and drive-shafts, if the propeller speed is to be 1/20th of the turbine speed.

Q.10.2(1). In the optimization of ship speeds, should the value of the cargo enter into the calculations?

Q.10.3(3). In a plant for desalinating seawater by forcing part of it through a membrane which does not allow the passage of salt, the seawater is raised to a pressure of 15 N mm^{-2}. The flow through the membrane is 0.40 of the flow into the plant, the remainder emerging as brine. To economize in energy consumption, the brine is passed out through a turbine, which helps to drive the seawater pump. If both pump and turbine have efficiencies of 0.80, what is the energy consumption per kilogram of fresh water? Can you suggest any means of reducing this figure, without increasing the pump and turbine efficiencies and keeping the yield of fresh water fixed at 0.40 of the seawater intake? Reflection about whether the pV term in the expression for enthalpy is 'real' energy may help.

Q.10.4. The hopper shown in Figure 10.1 is used for a material that may be regarded as a liquid of density 600 kg m^{-3} as far as the forces it exerts on the structure are concerned.

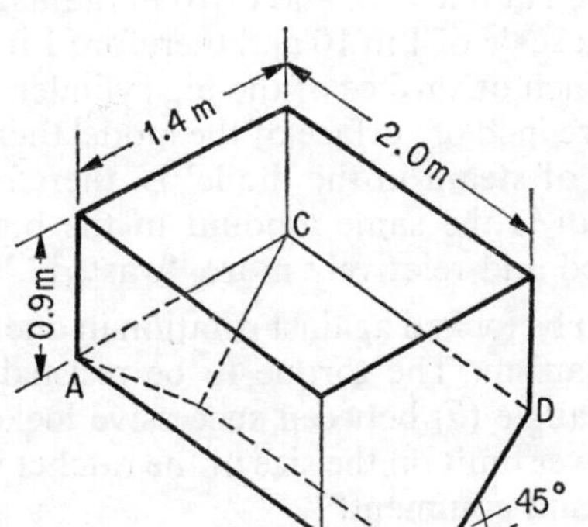

Figure 10.1 Hopper

The doors which close the bottom are hinged along AB and CD; the material is to be mild steel, with an allowable maximum shear stress of 60 N mm^{-2}.

(a) Estimate to ±20 per cent the weight of steel required using two- and three-tiered structures. Do not include the legs below ABCD.
(b) The doors are to close automatically after releasing the contents of the hopper. Suggest means of achieving this. Should the means be located at both ends? Try to make your arguments quantitative.

Q.10.5. In machine tools there are commonly several relative degrees of freedom between the work and the tool, some concerned with the generation of surfaces, one with the feed, and some with positioning. For example, in a lathe, when used for turning a cylinder, spindle rotation and sliding generate the cylindrical surface while the traverse supplies the feed. In a radial drill, rotation of the spindle and the feed generate the surface, while swinging the arm and moving the head along the arm give the two degrees of freedom necessary to position the axis of the hole. Try to find some general principles that are helpful in deciding which motions should be given to the work and which to the tool, and how the motions should be superposed, e.g. in a lathe, why put the cross-slide on top of the longitudinal slide and not the other way round?

Q.10.6. A silo for a military missile has a cover consisting of a horizontal slab of reinforced concrete 4.5 m square, weighing 200 tonnes. To fire the missile the slab must be moved 3 m to one side, and it is desirable to be able to cover and uncover the silo as quickly as possible. Analyse the problem and suggest suitable means. An electrical supply of 80 kW is available.

Q.10.7. Comment on the following passage from p. 36 of '*Steam Power in the 18th Century*' by Cardwell, which is about Watt's speculations on the poor performance of a miniature Newcomen engine.

'Now, Watt knew that since steam is condensed by cold metal surfaces, the greater the surface exposed the more the amount of steam condensed. Imagine a cylinder, open at one end, and of 10 in radius and 10 in long; also a "model" of it built to a scale of 1 in 10 and therefore 1 in radius and 1 in long. Then to every square inch of surface of the big cylinder there will be $3\frac{1}{3}$ in^3 of volume; to every square inch of surface of the model there will be only $\frac{1}{3}$ in^3 of volume. A cubic inch of steam in the model is, therefore, exposed to much more cold surface than is the same amount in the big cylinder. Relatively more will be condensed and relatively more "wasted".'

Q.10.8(2). A shaft is to be locked against rotation in one direction by a ratchet wheel and pawl mechanism. The torque to be resisted (T) is large and the maximum acceptable angle (θ) between successive locked positions is small. Show that there is a lower limit on the size of the ratchet wheel (i.e. the volume of the pitch cylinder) and comment.

Q.10.9(2). A gimbal ring with a mean diameter of 1 m is subject to loads normal to its plane at four equally spaced points, N,E,S,W. All the loads are of

100 kN, but those at N and S are oppositely directed to those at E and W. Estimate the maximum bending moment in the ring, and any other quantity relevant to a choice of cross-section. No heavy calculations are necessary, but symmetry considerations help.

Q.10.10(1). A meal is to consist of items X, Y, and Z and to contain at least 180 units of protein and 300 units of carbohydrate. What is the lowest-priced meal which meets these requirements?

	Units per new penny	
	Protein	*Carbohydrate*
X	5	4
Y	2	8
Z	6	3

Q.10.11(2). Figure 10.2 shows a portion of the tower of a suspension bridge looking north along the roadway. Due to a westerly wind loading, the tower

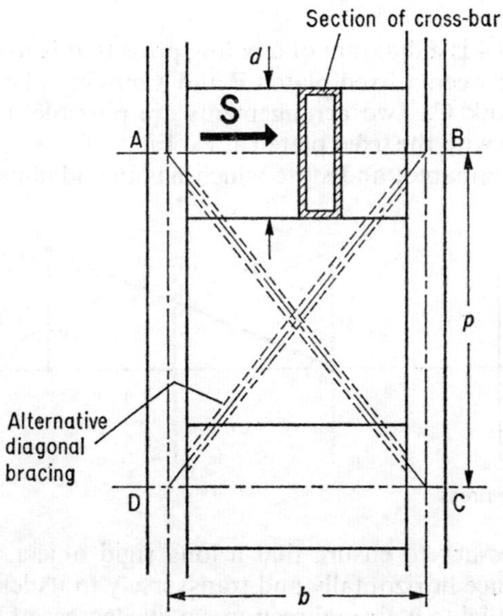

Figure 10.2 Suspension bridge tower

in the region shown is subject to a shear force S which may be regarded as constant over the bay. Estimate the volume of steel required in *one* cross-bar, using the maximum shear stress (Tresca) criterion with allowable *tensile* stress f. Compare this with the volume required for diagonal braces AC and BD, and find the ratio for $p = 30$ m, $b = 23$ m, $d = 11$ m.

Q.10.12. Figure 10.3 shows a mechanism for opening the doors of underground trains. Why is it impracticable in the form shown, and what is the least alteration that will make it workable? Comment on the matching.

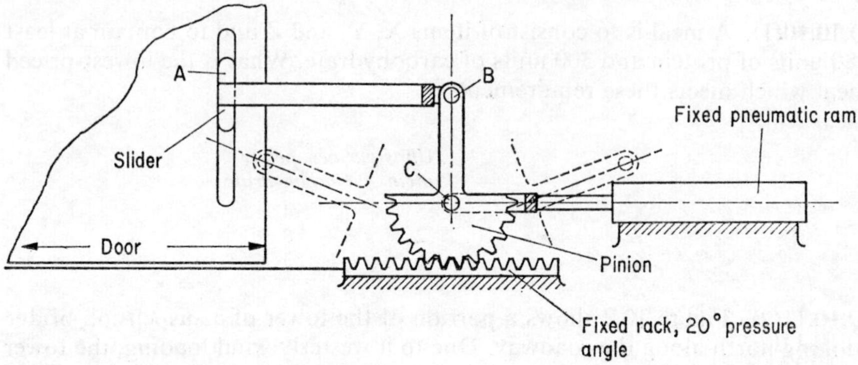

Figure 10.3 Sliding door mechanism

Q.10.13. Figure 10.4 is a diagram of a baling press that is to compress a loose fibrous material between a fixed platen P and a moving platen M by a single rotation of the crank C. Two arrangements are possible, (1) with the fixed platen at P_1 and (2) with the fixed platen at P_2. Find a few respects in which the schemes may be compared, and state which has the advantage in each case.

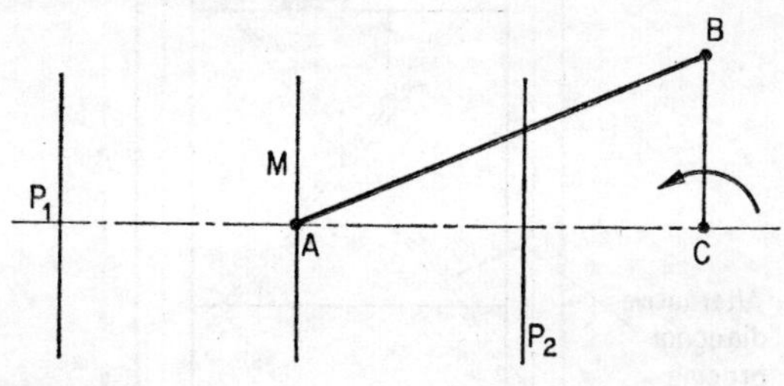

Figure 10.4 Baling press

Q.10.14. It is necessary to ensure that a long rigid object, which is free to move a small distance horizontally and transversely to its length at each end, should remain parallel to a given direction, i.e. its degrees of freedom must be reduced from two to one, of translation perpendicular to its length in the horizontal plane. Sketch kinematic devices to apply this kind of constraint. What figures of merit could you apply to your solution?

Q.10.15(3). A ship may be regarded for the purposes of this question as a rectangular steel tube filled with a solid cargo which makes the combined specific gravity 0.50 (relative to seawater), of which 0.42 is due to cargo. The depth of the hull is 20 m, and the centre of gravity of the ship and cargo is 10 m above the bottom.

(a) find the beam of the ship if the metacentric height is 1 metre,
(b) because it is the cheapest way of making a ship bigger, it is proposed to make it deeper, keeping the metacentric height at 1 m by fitting fixed ballast of specific gravity 7.5 relative to seawater in the ship's bottom. As a first step in the economic study, find by how many tonnes the ship's displacement will be increased for every tonne of extra cargo. If you can, find a treatment which keeps the algebra simple.

Q.10.16(3). A suction hovercraft travelling on a ceiling presents an almost insoluble control problem because it is inherently unstable—if it moves a little further away from the ceiling, the pressure in the lift pad will rise, and the vehicle will fall unless the exhauster can be stepped up in power very quickly. Yet a suction hovertrain is being developed in France.

Enunciate the geometric property of the lift pad that is essential for stability in this respect in both 'suck' and 'blow' hovercraft, and find a configuration for 'suck' hovercraft which has this property.

Q.10.17. An old puzzle consists of two square-section pieces of wood joined end to end with each face of the joint showing a dovetail (Figure 10.5). It is

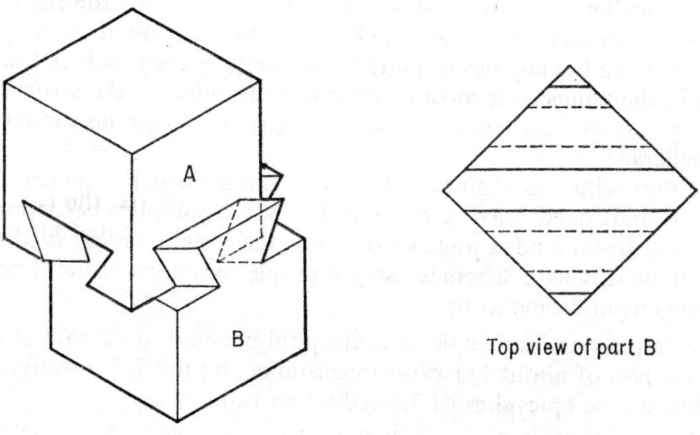

Figure 10.5

possible to make a similar puzzle consisting of two pieces of wood of equilateral triangular section which are joined together, with no voids in the interior, so as to show a dovetail on every face. How?

Q.10.18(3). A man climbs up out of a ravine mouth whose form is given by $z = x^2e^y$, where z is height and the units are hundreds of metres. He starts at the origin of co-ordinates and climbs always up the line of greatest slope to a plateau at $z = 4$. What is the width of the ravine where he emerges?

Q.10.19. Make a combinative study of ways of closing the end of a cylindrical vessel subject to internal pressure, and list the circumstances in which the various alternatives have advantages.

Q.10.20. To aid in manoeuvring them, some modern ships are pierced through from side to side by a horizontal tunnel containing a propeller. To drive such a propeller it is proposed to fit it with tip-jets ejecting water piped in through a hollow hub from the ship's fire-and-bilge pumps, which can pump seawater at approximately $0.7\ \text{N mm}^{-2}$. Comment on the matching problem.

Answers

A.10.1. If ω is the angular velocity of the propellers so that $20\ \omega$ is the gas-turbine speed, the interconnecting shafts should run at some intermediate speed, say, $k\omega$. If k is large, the torque will be small and the shafts light, except that critical-speed problems will limit this effect.

If we put the engines in line with the propellers, the maximum power to be handled by the interconnecting shafts is $\frac{3}{4}E$ (where E is the power of one engine) against E if the engines are elsewhere. The difference is greater than appears, in that the $\frac{3}{4}E$ is an emergency power while the E is a normal service requirement, and so there is a strong argument for putting the engines in line with the propellers. Can we find an argument for putting them elsewhere?

The engines in such an aircraft will be a sizeable fraction of the weight of the fuselage. Putting the engines in pairs at the wing tips will reduce the bending moment in the wings over most of their length, reducing the structural task, perhaps by one quarter compared with the case with engines directly behind the propellers.

Notice that with the engines in line with the propellers, the central interconnecting shaft need only be rated at $\frac{1}{2}E$. Some complicated arrangements can be made to take advantage of this, but the weight saving is likely to be very small, as it is with schemes using multiple meshings of bevel gears. The best arrangement seems to be:

(a) multiple pinion simple epicyclics dropping engine speed to about $7.5\ \omega$,
(b) bevel drives of about 1:1 ratio interconnecting the $7.5\ \omega$ shafts,
(c) further simple epicyclics of 7.5 reduction ratio.

A.10.2. Yes, as part of the capital cost.

A.10.2. 2.5 tonnes of seawater yield 1 tonne of fresh water, and the work done in pumping to $15 \times 10^6\ \text{N m}^{-2}$ is

$$1.25 \times 15 \times 10^6 \times 2.5 = 46.8 \times 10^6\text{J}$$

Of this, $0.8 \times 15 \times 10^6 \times 1.5 = 18 \times 10^6\text{J}$ is recovered in the turbine, leaving an energy consumption of

$$28.8 \times 10^6\text{J or 8 kWh per tonne of fresh water.}$$

Of this, the unavoidable work is that of forcing the fresh water through the membrane, or

$$1.25 \times 15 \times 10^6 \times 1 = 18.8 \times 10^6\text{J}$$

The remainder is due to pumping the other 1.5 tonnes into the high-pressure region and passing it out again through the turbine, and can be avoided simply by admitting the seawater and removing the brine at atmospheric pressure, i.e. by using a batch process, filling the plant at atmospheric pressure, pressurizing until the fresh water expelled is 0.40 of the original charge, and then depressurizing before refilling with seawater. There would be some power needed to compress the seawater and stretch the plant, which would not be negligible, and the fatigue condition on the plant would be expensive. The pressure over the membrane is roughly proportional to the salt concentration in the brine—you might try working out the optimum brine/fresh-water ratio k for a machine efficiency η $\{k = 1/\sqrt{(1-\eta^2)}\}$.

A.10.4. (a) The writer took 2 mm sheet for the lining of the two-tier version. At the depth of 1.5 m, the 'hydrostatic' pressure is

$$p = 1.5 \times 9.81 \times 600 \text{ N m}^{-2} = 0.0088 \text{ N mm}^{-2}$$

giving a bending moment in a span d encastered at each end of $\frac{1}{12}pd^2$, and this must be equal to or less than $fz = 120 \times 2^2\text{N}/6$.

Hence the allowable $d = 330$ mm, and from this may be deduced suitable pitchings for the reinforcements (which above the doors can with advantage be continuous horizontal 'rings'). The total weight of steel was about 300 kg. For a three-tier structure, the thinnest likely sheet was taken as 0.6 mm. With a flat sheet on the inside and a sheet with rectangular corrugations 20 mm deep on the outside spot-welded to it, reinforcements can be confined to the top, the level of the door hinges and the meeting edges of the doors. The weight of steel needed was estimated to be 210 kg. The material saving of perhaps £5 per hopper would almost certainly be lost in additional fabrication and finishing costs, and the two-tier design is probably cheapest, apart from the vulnerability of the lighter structure to corrosion.

(b) Springs or deadweights, with a retaining catch or some delay to ensure complete emptying.

If the doors are latched at one end only, the torque in a door can be calculated to be 9600 Nm. This can be carried out to the latch end by a tube 80 mm in outside diameter and 5 mm wall thickness, which will still have sufficient reserve of strength to act as a reinforcement in bending. The extra cost is likely to be less than £4 per hopper, which would probably be saved on the latching arrangements.

Another approach is to put two reinforcing arms on the door, say 600 mm in from the ends, and fit the single latch on the end of one of these arms. A smaller tube will then serve to take torque and bending moment (from the bending moment alone the ideal position of the arms is 410 mm from the ends—check this).

It may be worth considering making one door extend, say, three-quarters of the way across the plan view of the hopper, and replacing the other by a fixed slope which will now be inclined at 63.4° to the horizontal.

A.10.5. Important criteria are overall size, stiffness and the effects on accuracy. In some cases, the nesting order is dictated, e.g. generating movements must be 'inside' or 'farther from earth than' positioning movements. In deciding whether to move work or tool, it is usually better to move the tool for concave surfaces and the work for convex surfaces. For flat surfaces move the tool if the work is very large. It is usually better to nest motions of smaller travel inside those of larger travel. All these are guides only—each case must be treated on its merits.

A.10.6. Suppose 70 kW is available at the motor shaft and this is applied by a frictionless infinitely variable drive for the first 2.5 m. Then the kinetic energy is 70 000t J, where t is the elapsed time in seconds and so

$$70\,000t = \tfrac{1}{2} \times 200 \times 10^3 v^2,\; v = \frac{dx}{dt} = 0.84t^{\frac{1}{2}}$$

where v is the speed in metres per second, x is the travel in metres. Hence

$$x = 0.56t^{3/2},\; x = 2.5 \text{ when } t = 2.7 \text{ and } v = 1.38$$

If the 'eyelid' travels the last 0.5 m at an average speed of $\frac{1}{2} \times 1.38$ m/s under a uniform deceleration due to some buffer, the total time is

$$2.7 + \frac{0.5}{0.69} = 3.4\text{s (say)}$$

(We cannot quite reach this figure because we cannot start with an infinite acceleration.)

The problem can be eased by making the slab the mass of a spring-mass or pendulum-type vibratory system, with its 'at rest' position half-way between open and shut. If the natural frequency is 0.5 cycles per second, the opening and shutting times will be one second. To open the shut 'eyelid' we have only to release the catch which holds it shut. It automatically starts to vibrate, but is caught as soon as it comes to rest for the first time in the open position.

A steel coil spring working at a shear stress of 500 N mm^{-2} stores strain energy at a maximum density of 1.63×10^{-3}J mm^{-3} or 210 J kg^{-1} (section 4.5) with an efficiency of utilization of 0.5, which brings the average energy stored down to, say, 100 J kg^{-1}. The energy of a vibratory system is $\frac{1}{2} ma^2\omega^2$, where m is the inertia, a the amplitude and ω the angular natural frequency. Here $m = 200$ tonnes, $a = 1.5$ m and $\omega = \pi\text{s}^{-1}$, so the energy is 2.2×10^6 J, requiring springs weighing about 22 tonnes, which is rather discouraging. Air cylinders would be practicable, however. Mounting the slab on 'switch-back' rails would probably defeat its protective function.

In any 'vibratory' solution, the decay of amplitude could be overcome by biasing the 'position of rest' away from the current position of the slab, i.e. by increasing the pressure in the cylinders on the 'compressed' side. The size of compressor required would depend on the frequency of operation demanded.

A.10.7. More simply stated, the argument is that the mass of steam in the cylinder is proportional to L^3, while the area on which steam is condensing is proportional to L^2. This argument ignores the time factor— by the reasoning of section 4.7 the most suitable simple assumption is that velocities remain constant, so that the duration of the stroke is proportional to L and the product of the condensing area and the time it is exposed is proportional to L^3 and therefore the volume. In this case the little engine would be no worse off than the big one.

In more detail, the heat-transfer problem involves (a) the 'resistance' to heat transfer at the fluid/solid boundary and (b) the unsteady conduction in the wall. Three simple special cases can be distinguished, (1) where (a) is the bottleneck and the wall itself is to all intents and purposes an infinite heat sink, (2) where (a) is negligible so the inside face of the wall is always at steam temperature, and (3) where both (a) and (b) are negligible and the wall heats and cools right through on every stroke. Case (1) is the one we have just considered, and case (3) also puts the big and little engines on an equal footing. In case (2), however, the temperature (T) distribution through the wall will be somewhat as in Figure 10.6, where T_1 is the temperature at the end of the inlet

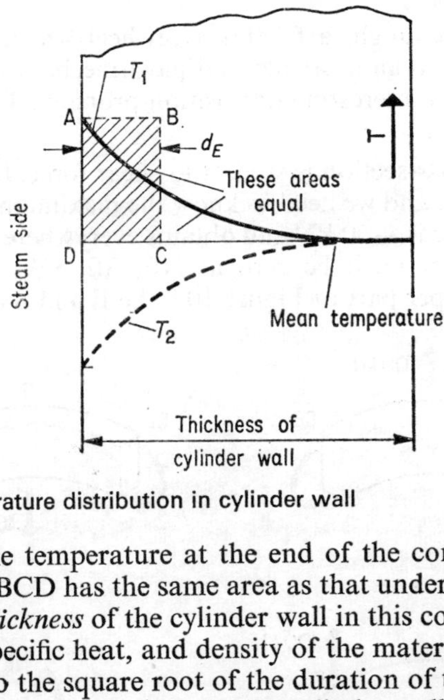

Figure 10.6 Temperature distribution in cylinder wall

stroke and T_2 is the temperature at the end of the condensing stroke. The shaded rectangle ABCD has the same area as that under the T_1 curve, so that d_E is the *effective thickness* of the cylinder wall in this context: d_E depends on the conductivity, specific heat, and density of the material of the wall, and is also proportional to the square root of the duration of the stroke, and hence to $L^{\frac{1}{2}}$. Case (2) reduces to case (3) with the cylinder wall thickness reduced to d_E, the cooling effect is proportional to the *effective* cylinder water equivalent and hence to $L^{5/2}$, and since 5/2 is less than 3, the big engine is better off than the little one: d_E is another useful concept.

$$d_E = \left(\frac{4kt}{\pi c\rho}\right)^{\frac{1}{2}}$$

where k is the thermal conductivity, c is the specific heat, ρ the density of the wall material and t is the duration of a stroke.

A.10.8. If we fix the pitch radius r of the ratchet wheel, the pitch of the teeth is θr and this fixes the size of the teeth and hence the load per unit facewidth they can carry at, say, $A\theta r$, where A is a constant. Now the load on the pawl, if we use only one, is T/r so that the facewidth is

$$\frac{T}{A\theta r^2}$$

and the pitch cylinder volume is

$$\frac{T}{A\theta r^2} \times \pi r^2 = \frac{\pi T}{A\theta}, \text{ which is fixed.}$$

If we use $2N$ pawls, the pitch cylinder volume is reduced to $\pi T/2NA\theta$. The best way will usually be to fit the pawls in diametrically opposed pairs (to balance the forces on the wheel), to increase the tooth pitch to $N\theta r$, and to stagger the N pairs of pawls so that successive pairs engage at angular intervals θ.

When θ is small enough, a friction freewheel which makes θ 'infinitely' small will be smaller than a ratchet and pawl mechanism. The sprag type is most compact and an interesting disposition problem. The analysis is similar to that of section 4.9.

A.10.9. The ring cross-section is subject to shear force, bending moment and also possibly torsion, and we need to know the maximum values of these. The maximum shear force is ± 50 kN and obtains everywhere except under a load. The bending moment must be zero at NE, SE, SW, NW for reasons of symmetry. In the upper part of Figure 10.7 the E and S quadrants have been

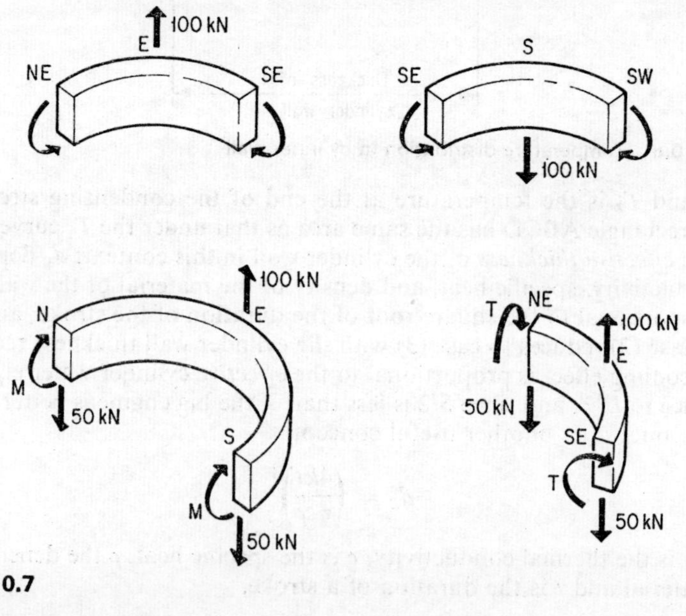

Figure 10.7

shown with bending moments at the ends, and it is clear that if the sign of the moments is correct in one it must be wrong in the other—the only possibility is that the moments are zero. The maximum bending moments must occur under the loads, and can be found by the equilibrium of the eastern half ring (Figure 10.7)—the maximum M is 25 kN m. The maximum torque occurs at NE, SE, etc. and can be found from the equilibrium of the eastern quadrant (Figure 10.5) by taking moments about the NE-SE axis, giving

$$2/\sqrt{(2)}T = 100 \times 0.5\left(1 - \frac{1}{\sqrt{2}}\right) \text{ kN m, } T = 10.4 \text{ kN m.}$$

Feilden records a case [37] of this torque being overlooked, an error which would not occur if Torroja's precepts were followed (section 10.4).

A.10.10. 30p worth of y, 20p worth of z

If x, y, z pence are spent on X, Y, Z, the condition that there is just enough protein is

$$5x + 2y + 6z = 180$$

which is a plane dividing the *feasible* from the *non-feasible* region of x, y, z space. Similarly, from the carbohydrate requirement,

$$4x + 8y + 3z = 300$$

One way of finding the answer is to draw the traces of these planes in the xy, yz, zx planes, and also the traces of planes of constant cost ($x + y + z = C_T$).

A.10.11. There will be zero bending moment in the legs halfway between crossbars and the shear in each leg is $S/2$. The bending moment in the ends of the crossbars is $\frac{1}{2}pS$, and the shear force in them is pS/b. The combined area of the flanges must be

$$2 \times \frac{\frac{1}{2}pS}{fd} \text{ and that of the shear webs } \frac{2pS}{bf}$$

The total volume of material in one crossbar is thus

$$\frac{Sp}{f}\left[\frac{b}{d} + 2\right]$$

A pair of diagonal braces will be of length $(b^2 + p^2)^{\frac{1}{2}}$ and carry a combined load

$$S\frac{(b^2 + p^2)^{\frac{1}{2}}}{b}$$

The volume of material required is thus

$$\frac{S}{f} \cdot \frac{b^2 + p^2}{b}$$

so the ratio of volumes is

$$\frac{bp\left(\frac{b}{d}+2\right)}{b^2+p^2} = 1.98$$

for the given dimensions. It is possible to graduate the thickness of plating in the flanges of the crossbars and reduce this ratio. However, the crossbars also induce bending moments in the legs which have to be paid for as well and add to the functions competing for a share of the allowable stress.

Notice that even to secure this diseconomy the crossbars must be more than half the depth of the clear space between and are much the most heavy-looking part—quite out of scale with the slender grace of the rest of the bridge and performing, for all their ponderousness, a relatively small structural task.

A.10.12. To avoid jamming at A the slider would have to be either very long or frictionless: the first is awkward to accommodate and the second expensive. A pivot rather than a slider at A solves the problem. Also, some sort of crosshead at C seems desirable to take the side thrust from the gear teeth.

Since it provides an increase of mechanical advantage at both ends of the travel, where it may be required, the matching seems good.

A.10.13. Compactness: scheme 2 is clearly more compact,
Complexity: the connecting rod and crank must be duplicated in scheme 2,
Structural task, static parts: the static parts carrying the pressing force have perhaps half the structural task in scheme 2,
Matching: the pressing action can occupy a greater fraction of a turn of the crank in 2, giving a smaller maximum torque.

A.10.14. Figure 10.8 shows three possibilities. Of these (a) uses material most economically and (b) is the tidiest. Little can be said in favour of (c), but if the inner arms of the two similar levers are made much longer than the outer ones reasonable structural economy is possible. (Why is this the best proportion for the levers?—see A.5.11.)

In this sort of problem the movement to be accommodated is generally small, but on the other hand the parallelism required must be correspondingly good. With large loads to be moved, high stiffness is usually required, and here the rather clumsy version (a) wins. As a figure of merit, we could use

$$\frac{\text{angular stiffness, } k \text{ (say, in Nm/rad)}}{V\text{, volume of material used for linkage}}$$

It is interesting that if we try to make this non-dimensional we do not obtain a logical form, since k/EV is non-dimensional but does not involve the length AB (Figure 10.7) over which parallelism is to be maintained, which makes nonsense of it. The reason for this anomaly is that the structural task (section 8.1) is nil. In the mechanism of Figure 10.8(a), for example, all the members can be vanishingly small except for EF. Now by the argument of

A.5.11, EF can also be made as thin as we wish while keeping its effective stiffness, simply by gearing its motion up—the function it performs is analogous with that of the brake cable. While this is not practical, it does show that there is no *absolute* lower bound for V for a given k. See reference 50 for examples of this sort of problem.

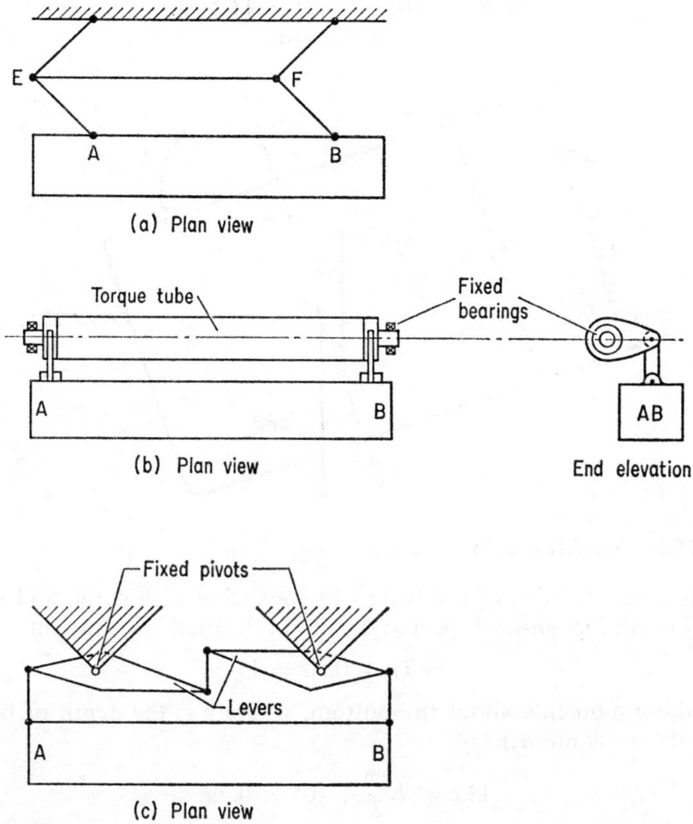

Figure 10.8

A.10.15. (a) If b is the beam, d the depth, ρ the density of seawater, then per unit length of ship the displacement is $\frac{1}{2}\rho bd$. The centre of buoyancy B is $d/4$ below the centre of gravity G, so that when the ship is tilted through angle θ, there is an overturning moment per unit length of $\frac{1}{8}\rho gbd^2\theta$ (Figure 10.9). The restoring moment of the water surface is $(1/12)\rho gb^3\theta$, so that

$$\begin{aligned}\tfrac{1}{12}\rho gb^3\theta - \tfrac{1}{8}\rho gbd^2\theta &= \text{displacement} \times \text{metacentric height} \times \theta \\ &= \tfrac{1}{2}\rho gbd\theta \times \text{metacentric height.}\end{aligned}$$

(In future equations of this form we shall leave out the factor $\rho gb\theta$.) Putting $d = 20$ m, metacentric height $= 1$ m, we find $b = 26.8$ m.

(b) 2.7. This is a differential study, but it is difficult by differentiation. Use actual figures, and start by assuming a given *increase in draught* to make the algebra simple. If the draught is 11 m, and the centre of gravity is now x m above the bottom, we can find x directly from the condition that the metacentric height is 1 m.

$$\tfrac{1}{12}b^2 - 11[x - 5.5] = 11 \times 1,$$
$$x = 9.95 \text{ m}$$

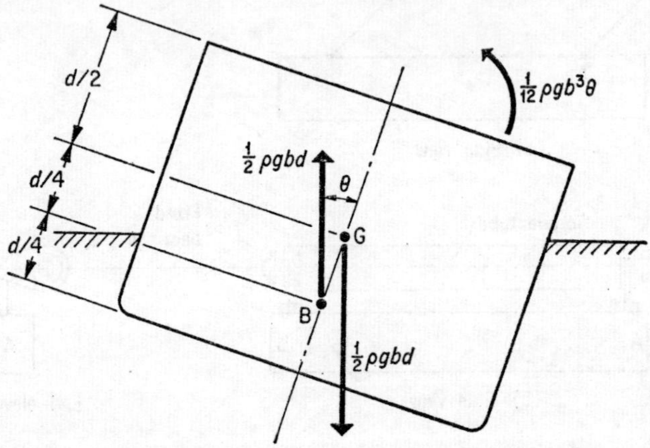

Figure 10.9 Stability of ship

We now use the idea of the ballast as coexisting with cargo and steel but having just an *incremental* specific gravity of 7. Then the draught

$$= 7y + 0.5d = 11 \qquad (1)$$

Also, taking moments about the bottom, where y is the depth of ballast in metres, the mass moment is

$$11x = 7\frac{y^2}{2} + 0.5 \times 0.5d^2$$

From equation 2, ignoring the y^2 term, $d = 20.95$ m. From equation 1 $y = 0.075$ and substitution in equation 2 gives a negligible change in d. (This is much better than substituting and solving quadratics.) Hence answer—note additional draught due to cargo $\simeq 0.42(0.95–0.075)$ m.

Now as the displacement increases by nearly 3 tonnes for every tonne of additional cargo, the increase of propulsion costs (which are about one third of the whole—section 3.3) is nearly equal to the costs of transporting one tonne in the original ship. When we consider also the cost of the ballast (which might be pig-iron at, say, £25 per tonne) the case against the 'deep ship' looks complete. If we use concrete at perhaps one-tenth of the price, the ratio of increase of displacement to increase of cargo becomes even more unfavourable. On the other hand, we have not taken account of the reduced thickness of deck and bottom plating possible because of the increased *d*. We can conclude that the pig-iron-ballasted ship is not viable, but the concrete-ballasted one may be a 'runner' and warrants a more careful study, taking account of double bottoms, the ends of the ship, etc.

A.10.16. Round more than half the periphery of the lifting pad the lifting pad side of the gap between pad and support must lie above the support. Imagine a suction hovertrain running on the bottom flange of an I-beam, using a rectangular lifting pad which curls over the flange at the sides so that the gap along the sides has the pad above the support (Figure 10.10). Then if the pad is substantially longer than it is wide, the condition for stability will be met.

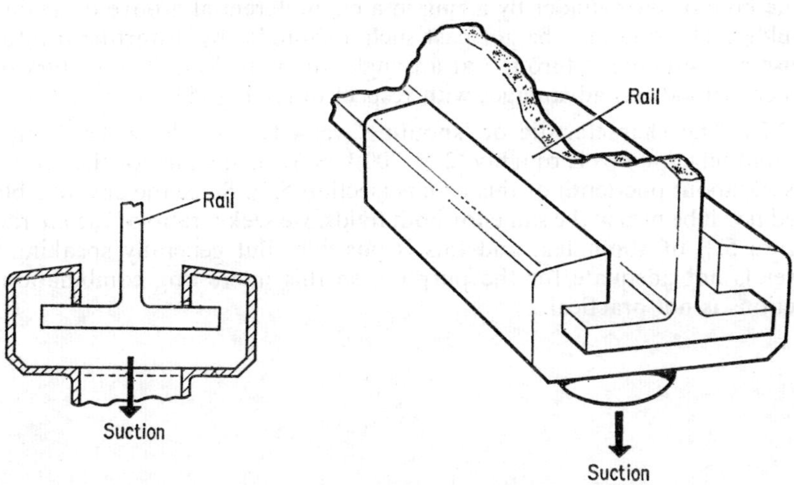

Figure 10.10 Pad configuration for stable suction hovercraft

A.10.17. There are two solutions at least. Section 2.3 gives the key—one solution is to make the joint a turning pair with the centre line of one side as axis. The other solution is a combined sliding-convergent joint, and is rather less obvious. There may well be other solutions the writer has not found.

A.10.18. Slightly over 300 m. The contours are given by $z = \text{const.} = x^2e^y$. Differentiating with respect to x for constant z gives

$$2xe^y + x^2e^y\frac{dy}{dx} = 0, \text{ or } \frac{dy}{dx} = -\frac{2}{x}$$

Now on a contour map with y, x axes on it, the slope of the contours is $-2/x$ so that the slope of the line of steepest slope, which is at right angles to the contours, is $x/2$, and the differential equation of the man's path on the map is $dy/dx = x/2$, or $y = x^2/4$ since he starts at the origin. He arrives at the plateau at the point

$$z = 4 = x^2e^y = 4ye^y$$

Solving (by successive substitutions, say) $y = 0.568$ (roughly) $x = 1.5+$.

A.10.19. Useful classifying principles are:

(a) extra parts or not,

(b) number of surfaces in shear (clearly, there must be at least two).

Cases of no extra parts, besides welded, brazed and simple screwed joints, are the interrupted, stepped, screw threads (Welin screw) used for the breaches of big guns and the transversely sliding breach blocks of much artillery. Then there is the Welsh plug, a dished disc of material expanded by flattening into a circumferential groove in the bore of the cylinder.

Important cases of two surfaces in shear are those where a plug is retained in the bore of the cylinder by a ring in a circumferential groove or against a shoulder. The ring may be got past such a shoulder by distorting it into an ellipse and slipping it through at an angle, or by making it in sectors—the former method has advantages with respect to sealing [53].

A.10.20. The characteristic or 'spouting' velocity c of the water from the fire- and bilge-pumps is roughly $(2 \times 700)^{\frac{1}{2}} = 37$ m/s, while for the propeller we seek about one-tenth of this or less (section 5.2). Since the effective blade speed u will be nearly the same for both fluids, we seek a ratio of the u/c ratios (Figure 5.2) of about ten, and this is possible. But generally speaking the power is not adequate for the purpose, so this interesting combination of functions is not practical.

References

1 EUSEC Conference, September 1966.
2 GHISELIN, B., *The Creative Process* (New American Library, 1952).
3 RUHEMANN, M., *The Separation of Gases* (Oxford, 1949).
4 JACKSON, R. G., and FFOOKES, R. C., 'Sea transportation of L.N.G.' Le petrole et la mer, III, No. 302, International Congress of Petroleum and the Sea, Monte Carlo, 1965.
5 JUNG, I. *Swedish Marine Turbine and Gear Development* (Society of Naval Arch. and Mar. Engs. N.Y. 1966) Spring Meeting, Paper 6.
6 WANKEL, F., *Rotary Piston Machines* (Iliffe, London, 1965).
7 HARARY, F., JESSOP, N., LUCKMAN, J., and SPRINGER, J., *Nature*, 1965, **206**, No. 4979.
8 ALEXANDER, C., *Notes on the synthesis of form* (Ph.D. thesis, Harvard, 1962).
9 CALDWELL, D. S. L., *Steam Power in the 18th Century* (Sheed and Ward, London, 1963).
10 'Stirling Machine Developments,' *The Engineer*, Nov. 26, 1965, pp. 906–12.
11 KAYS, W., and LONDON, A. L., *Compact Heat Exchangers* (McGraw-Hill).
12 BATTERSBY, *Mathematics in Management* (Pelican, London, 1966).
13 GASS, S., *Linear Programming: Methods and Applications* (McGraw-Hill).
14 FOX, C., *An Introduction to the Calculus of Variations* (Oxford, 1950).
15 DICKINSON, H. W., *A Short History of the Steam Engine* (Cambridge, 1939), p. 188.
16 GIFFORD, E. W. H., 'The Development of Long-Span Prestressed-Concrete Bridges,' *Struct. Eng.*, Oct. 1962, p. 325.
17 COX, H. L., *The Design of Structures of Least Weight* (Pergamon, 1965).
18 MACMILLAN, R. H., and DAVIES, P. B., 'Analytical Study of Systems for Power Transmission,' *J. Mech. Engng. Sci.*, 1965, (No. 1) **40**, London.
19 SMEED, R. J., 'A Theoretical Model of Commuter Traffic in Towns,' *J. Inst. Maths Applic.*, **1**, pp. 208–25.
20 DAVIES, W. J., 'Performance and Reliability of High-Duty Gearing for Aircraft,' Paper 16, Proc. Intern. Conf. on Gearing 1958, I.Mech.E., London.
21 HAUSEN, H., 'Verlustfreie Zerlegung von Gasgemischen durch umkehrbare Rectifikation,' *Z. techn. Phys.*, 1932, **6**, pp. 271–7.
22 LJUNGSTROM, F., *Proc. Inst. Mech. E.*, 1949, **160**, 216.
23 STODOLA, A., *Steam and Gas Turbine* (P. Smith, Massachusetts).
24 PAYNE-GALLWEY, SIR R., *Missile-throwing weapons of the ancients.*
25 HOMER, Odyssey, Book 21.
26 FRENCH, M. J., 'A Fundamental Limitation of Mechanisms,' *Mech. Eng. News*, Feb. 1969, American Society Engineering Education, Oklahoma State Univ.
27 GRYLLS, S. H., 'The History of a Dimension,' *Proc. Inst. Mech. Eng.*, 1963, **178**, Pt. 2a, p. 1.
28 TUPLIN, W. A., *Gear Load Capacity* (Pitman, 1961).
29 FRENCH, M. J., 'Gear Conformity and Load Capacity,' *Proc. Inst. Mech. Eng.* 1965–66, **180**, Pt. 1.
30 GLEGG, G. L., *The Design of Design* (Cambridge Univ. Press, 1969).
31 BITTER, F., *Rev. Sci. Inst.*, 1940, **11**, 373.
32 LAITHWAITE, E. R., *Induction Machines for Special Purposes* (London, 1966).

33 PARKES, L. R., *Design for a Marine Diesel Engine*, British Welding Research Assoc. Bulletin Vol. 6, No. 4, April 1965.

34 NIEMANN, G., *Maschinenelemente*, Vol. 2 (Springer, Berlin, 1960).

35 BOROVICH, L. S., *Lowering the Weight of Reduction Units by a Rational Choice of Basic Parameters, Increasing the Loading of Gearing and Decreasing its Weight*, ed. M. M. SAVERIN (Pergamon).

36 JOUGHIN, J. H., 'Naval Gearing—Wartime Experience and Present Developments,' *Proc. Inst. Mech. E.*, 1951, **164.**

37 FEILDEN, G. B. R., 'A Critical Approach to Design in Mechanical Engineering,' Bulleid Memorial Lectures, 1959, University of Nottingham.

38 TUPLIN, W. A., 'Improved Mousetrap,' *The Engineer*, April 30, 1965, p. 773.

39 OSBORNE, A. F., *Applied Imagination* (Scribners, 1963).

40 DIXON, J. R., *Design Engineering—Inventiveness, Analysis and Decision* (McGraw-Hill).

41 ATKINS, A. J., and ELLIS, F. E., 'An Automatic Transmission for Small Cars,' *Proc. Inst. Mech. Eng.*, 1966–1967, **181**, Pt. 2A, No. 6.

42 I.C.I. Limited, *Assessing Projects* (Methuen).

43 BAUMOL, W. J., *Economic Theory and Operations Analysis* (Prentice Hall).

44 MATOUSEK, R., *Engineering Design* (Blackie).

45 RODENACKER, W., *Methodisches Konstruieren* (Springer, Berlin, 1970).

46 MARPLES, D. L., *The Decisions of Engineering Design* (Institute of Engineering Designers, London, 1960).

47 REULEAUX, F., *Kinematics of Machinery* (Dover).

48 JONES, J. C., 'Design Methods Compared,' *Design*, Council of Industrial Design, London, Aug. and Sept. 1966.

49 TORROJA, E., *Philosophy of Structures* (Univ. of California Press, Berkeley, California, 1967).

50 McLEOD, R. C., 'Practical Turbinology,' *The Engineer*, April 27, 1962, p. 728; Aug. 3, 1962, p. 188; April 12, 1963, p. 667; June 28, 1963, p. 1144; Nov. 15, 1966, p. 800.

51 BOOKER, P. J., *Principles and Precepts in Engineering Design* (Institution of Engineering Designers, London, Sept. 1962).

52 D'ARCY THOMPSON, W., *On Growth and Form*, abridged version edited by J. T. BONNER (Cambridge, 1961).

53 POLYA, G., *How to Solve it* (Doubleday, New York, 1957).

54 ANDERSON, H. H., 'Self-Seal Joint for Very High Pressures,' A.S.M.E. paper No. 5, Pressure Vessels and Piping Div. Winter meeting, 1967.

Index

ABSTRACTION, benefits of, 5, 8, 103, 199
Adaptive mechanism, 128, 129
Ad hoc measure, 86
A.E.I., 143
Air cylinder, 126
Air separation, 8, 55
Aircraft, man-powered, 104
Allowable stress, 63, 155
Alternatives, search for, 203
Alternator rotor, 141
Analysis of Interconnected Decision Areas (A.I.D.A.), 28
Animals, respiration of, 96
Approach, additive, 22
  diversifying, 4
  fresh, 206
Arbitrary decision, 149, 173, 198, 208, 209
Area/volume ratio, 145
Assets, 191
  exploitation of, 115, 181, 191
Austenitic and ferritic steel, 198
Automatic transmission, 92, 207
Autonomous local optimizing principle (A.L.O.P.), 66, 85
Availability, 89, 121
Available energy, 89, 121, 204
Axial flow compressor, 12, 81

BACKLASH, 178
Balance wheel, 104
Baling press, 218
Ball joint, 158
Balloon, 182
Beam, 9, 71, 75, 146
  of minimum weight, 90, 190
  redundant, 181
Bearings, 90, 167
  plain, 90, 196
  tilting pad, 165, 176
Bed, 200
Bedroom, 157
Bending strength, 154, 158
Blade fixings, 12, 137
  propeller, 110
  turbine, 81, 94, 110, 198
Bolts, friction grip, 40
  prestressed, 161
Bomber, 80
Bounds, 90
Bows, long and reflex, 124
Brachistochrone, 62, 85, 150
Brainstorming, 206
Brunel, I. K., 215
Building bent, 187
Bull-wheel, 193

CALIPER, brake, 127, 139
Car, 28, 33, 92, 160
  suspension, 208
Carnot's theorem, 91
Centring devices, 41, 178
Changes to be followed through, 46, 194
Check lists, 196, 211
Clamp, 75
Classification, 11, 23
Coal mining, 61
Coin problem, 105
Compliant member, 172
Computer, 3, 8, 30, 61, 78
Condensation of lines in tables, 31, 33
Configurations, 19
Connecting rod, 158
Contact stress, 98, 153
Contrarotating propellers, 25
Cooling air shroud, 78
Corner, power, 9, 92
  work, 125, 127

Costs, 48, 50, 61, 144, 148
of gearing, 23, 191, 195
of propulsion, 45, 114
running and capital, 44, 56, 212
Coupling, gear tooth, 184
Cox, H. L., 214
Crane, level-luffing, 131
vehicle-mounted, 133
Criterion, 3, 155
maximum shear stress, 63, 86
Tresca, 63, 86
Cryogenic plant, 130
Cupboard, 157
'Cut-price' solutions, 38, 128
Cylinder, closure, 219
thickwalled, 133

D'Arcy thompson, W., 214
Decimalization of currency, 208
Decision, area, 29
tree, 213
Degrees of freedom, in machine tools, 216
of choice, *see* Freedom of choice
spatial, 163, 185, 207
to balance loads, 165, 167, 176
Desalination, 215
Design, anatomy of, 1
automatic, 85
conceptual, 3
cost of, 212
elastic, 163
kinematic, 163
methods, 4, 7
of gears, 96
of gear teeth, 152
of living creatures, 214
of minimum weight, 65
philosophies, 4
problems, 11
process, 85
rationales, 4
study, 32, 84
thermodynamic process, 8
Detailing, 3
Diaphragm, 75, 172
Differential, gear, 92
study, 32, 55, 81, 84, 149, 228
Dimensional analysis, 97
Dimensionless groups, 98
Disc, 13, 32, 78, 95, 198
brake, 38, 126, 139, 165
centring, 41, 138
of optimum design, 63
Disposition, 137, 190, 224
Distillation column, 102
Diversification of approach, 6
Double-skinned construction, 148
Dovetail, fixing, 12, 137
puzzle, 219
Drum, 13, 16, 32
Ducts, hinges in, 180, 183

Ecology, 214
Effective, cross-sectional area, 82
hoop strength, 70
thickness, 223
Efficiency, Froude, 109
of joint, 138
of use, 87, 117
thermodynamic, 88
volumetric, 151
Elastic, compared with kinematic, 177
design, 163, 171
elements, 171, 174, 176
foundation, 200
hub design, 66
pairs, 171
Electric cooker, 205
Electrical, machine, 95, 141
power source, 130
Electromagnets, 156
Elimination, 21, 25, 30, 39
Energy storage, 119
Engagement, sliding, 13
Epicyclic gearing, 166
compound planet, 170
load-sharing, 185
multiple meshing, 194
spider, 167, 170, 172
tooth numbers, 208
Euclidean geometry, 4
Evaluation, 2, 212
Evolutionary techniques, 34
Exhausted, exhaustible, 10, 19

Failure, fatigue, 153
of discs, 69
of fixings, 138

Fatigue, 153, 156
Feasible area, 57, 225
Feilden, G. B. R., 200, 214, 225
Figure of merit, 86, 103, 147, 153, 189
Fir-tree fixing, 12, 138, 198
Flanged joint, 200
Flexibility, 209
Floor of house, 146
Flywheel, 104, 126, 130
Foil-sampling machine, 23, 25
Frame, of diesel engine, 182
  redundant, 181
Freedom of choice (degrees of), 10, 28, 29, 153, 198, 200
Freewheel, 224
Fuel, boiler, 115
Function, 11
  analysis, 11, 12, 159, 160
  auxiliary, 7, 39
  combination of, 22
  redistribution of, 35
  separation of, 35, 162

GAS-COOLED REACTOR, 120
  hinges in ducts, 180, 183
  matching of circulators, 133
  pressure vessel, 35, 196
Gas turbine, 53, 78, 94, 138, 198
Gear casings, 192
  grinder, 104
Gearing, automatic transmission, 207
  design, 96
  epicyclic, *see* Epicyclic gearing
  for contraprops, 194
  involute, 152
  non-involute, 155, 193
  of turbine, 110
  reduction, 23, 110, 114
  root design, 158
  strength, 153
  tasks and values, 191
  tooth design, 152
  tooth value, 193
  variable speed, 92
  Wildhaber–Novikov, 193
Geometries of elastic elements, 175
Gimbal ring, 216

HANDBRAKE, 126, 134
Heat, elements, 56
  engine, 158
  exchanger, 50, 71, 120
  flow in rod, 96
  pipe, 204
Heating and cooling curves, 120, 132
Helicopter engine mounting, 164, 181
Hertz's problem, 98
Hill-climbing, 60, 219
'Holding-together power', 9, 64, 93, 190, 196
Hole, weight of, 82
Hovercraft pad, 219
Hub, 66, 79, 82
Hull, 19, 77
  structure of, 143
Hybrid, 19, 23, 34
Hydraulic transmission, 92
Hydrostatic, bearings, 168
  transmission, 93
Hysteresis, 177

INDUSTRIAL designer, 212
Insects, respiration of, 96
Insight, 4, 5, 74
Inspiration, 5, 7
Instruments, 180
Insulation, economic thickness, 143
Interaction, 25
Internal combustion engine, 94, 114, 137, 200
  medium speed, 115
Inversion, 141, 204

JOINT, 12, 229

KELVIN, 163
Kernel table, 32
Kinematic, compared with elastic means, 177
  condition, 101, 170
  design, 163
  pairs, 13
  slide, 177
Knights, problem of, 6

LAGRANGIAN multiplier, 62, 159
Lamp, Terry 'Anglepoise', 132
Lathe, 183, 216
Layshafts, multiple, 178
Letter balance, 130

Level-luffing crane, 131
Limits, 90
Linde–Fränkl process, 8
Liquefaction, of gas, 132
 of natural gas, 124
 of nitrogen, 89, 190
Liquid natural gas (L.N.G.) tanker, alternative orders, 19
 design of, 143, 147
 economic speed, 46
 hybrid tanks, 34
 ice on hull, 6
Liquid natural gas, cold in, 204
Liquid nitrogen, manufacture, 89, 190
 return cargo, 205
 tank, 88
Ljungstrom, F., 107
Load, diffusion, 32, 37, 64, 182
 distribution on roller, 180
 sharing in gears, 178, 185
 statically determinate, 164
Logical chain, 141, 149, 204

MAP, 6
Marples, D. L., 213
Matching, 70, 108, 226
 circulating compressors, 133
 gear teeth, 171
 in single machine, 112
 in thermodynamic processes, 120
 mainsheet, sailing canoe, 133
 of spring, 117
 windlass, 108
Matousek, R., 213
Meal, balanced, 217
Mechanism, door opening, 218
 for pure translation, 218
 level-luffing, 131
 ratchet wheel and pawl, 216
 to apply pure torque, 105
Membrane tank, 22
Metacentric height, 219
Missile, silo for, 216
Mixed refrigerants, 124
Motor mower, 41
Moulton bicycle, 198

NEED, 1
Newcomen engine, 9, 35, 216
Newton–Raphson method, 152
Niche, ecological, 2, 214
Nuclear reactor, 35, 120
 see also Gas-cooled reactor

'ONE-AND-A-HALF power law', 37, 94
Operational theory, 200
Optimization, 44, 137, 140
 classification of, 58
 of shaft diameter, 81
 with more than one degree of freedom, 49
Optimizing principle, autonomous local (A.L.O.P.) 66, 85
Optimum, sag/span ratio of bridge, 47
 shaft, 83
 speed of ship, 46
 structure, 85
 virtual shaft, 84
Options, graph, 29
 pivotal, 33
 table of, 11, 14, 31, 33
 viable, 14

PARAMETERS, 196
 in coal mining, 61
Parameter-space, 19
Parametric mapping, 14, 30, 41
 of turbomachinery, 111
Philips piston seal, 36
Pigeon hole, 11, 25
Pipe flow, 71, 98
Pivot, crossed-spring, 177
Plastic design, 68
Plate, stiffened, 145
Pram, 41
Pressure angle, 153
Pressure vessel, 9, 93, 104, 133
Prestress, 125, 161, 196
Problem, analysis of, 2
 programming, 56, 225
 statement of, 2
 turning value, 59
Programming problems, 56, 225
Propeller, 109
 manoeuvring, 220
Proportion, 99
Pump, swashplate, 71
Pumping problem, 71

QUANTIFYING, benefits of, 8, 30, 190
'Quick sum', 80

RADAR, 81
Rankine cycle, 158, 204
Refrigerator, 55, 88, 90
  absorption, 101
  mixed refrigerant, 124
Reinforcement, hub, 82
  pressure-vessel, 196
Repêchage, 32, 39
Repertoire, engineering, 12, 111, 163, 167, 177, 203
Reuleaux, F., 213
Ring, 13, 32, 64, 78
  gimbal, 216
  spring, 87
Rocket, ELDO, 182
Rodenacker, W. G., 23, 213
Roller, bearing, 99, 180
  contact stress, 98
Rolls–Royce, 15, 18, 20, 168, 178, 199
Rotor, alternator, 141
  axial flow compressor, 12, 81, 196
  minimum weight, 85
  solid, 16

SAG/SPAN ratio of suspension bridge, 47
Sailing canoe, 133
Scale effects, 93
Scales, bathroom, 100
Scaling, 93
  of electrical machine, 95
  of engine, 94
  of turbine, 94
Scantlings, 145, 181
Schemes, 1
  embodiment of, 3
Scuffing, 155
Second Law of Thermodynamics, 91
Self-aligning, bearing, 167
  planet, 172
Sensible heat, 9
Shaft, 14
  optimization, 81, 103
  quill, 171, 178
  spin drier, 104, 191
  splined, 138
  virtual, 14, 18, 84
Ship, 45, 70, 76
  propulsion, 108, 114
  structure, 145
Specific, speed, 115
  velocity, 109
Spider, 172
Split-drive transmission, 93
Spokes, 32, 64
Spring, 86, 161, 177, 222
  compression, 86
  fluid, 104
  leaf, 87
  non-linear, 120
  rear, of car, 208
  ring, 88
Stability, 151, 218
Statical condition, 100
Steam-driven water pump, 125
Steam turbine, 81, 105, 109, 114
  compounding, 110
  gearing, 110
  multiple exhaust, 113
  radial flow, 114
Steepest descent, 60
Step, 7
  inventive, 4, 8
  mental, 4
Stirling cycle, 9, 36
Stool, three-legged, 163
Strain energy, 86, 87, 222
Strategy graph, 29
Structural, economy, 180, 226
  function, 21, 190
  task, 21, 33, 34, 39, 226
Structure, alternatives, 64, 144, 145
  hopper, 215
  loading coefficient, 147, 182
  minimum weight, 91
  optimum, 85
  ship, 145
  tiered, 147, 216
Strut, 104
Submarine, deep sea, 41
  magnetohydrodynamic, 103
Sub-problems, 11
Surface, plate, 165
  strength, 153
Suspension bridge, 47, 107, 155, 217
Swashplate pump, 71
Synthesis, 22, 32

Synthesizing, approach, 33
principle, 4
Systematic Design Method of Rodenacker, 23

TABLES, kernel, 32
of options, 11, 14, 31, 33
Tanker, 45, 76
L.N.G., *see* Liquid natural gas tanker
Task, 103, 190
structural, 21, 33, 34, 39, 64, 220
Task/cost ratio, 58, 190
Temperature mismatch, 123
Tendons, reinforcing, 196
Thermal, conductivity, 204
stresses, 79
Thermodynamic process, design, 8
matching, 120, 123
Throttling, equivalent process, 123
Tolerances, costs of, 56
effects of, 171, 175, 221
Torroja, E., 9, 225
Torsion rod, 87, 175
Trading off, 93, 143, 193
Traffic, 105
Tresca criterion, 63, 175
Turbine, Francis, 116
gas, *see* Gas turbine
helium, 130
impulse, 110
Pelton, 112, 116
steam, *see* Steam turbine
water, 116
Turbo-alternators, 112
Turbomachinery, 110
Typewriter, I.B.M., 39

UNDERCARRIAGE, 208

VALUE, 27, 190
engineering, 190
in disposition problems, 139
Valve-operating gear, 41
Variable-speed gear, 92
V.T.O.L. aircraft, 215
Vibration, 12, 85
Viewpoint, change of, 100
da Vinci, sketches, 215

WANKEL, F., 23, 213
engine, 23
Watt, 9, 35, 81, 216
Wavelength, 202
Web, 79
Welin screw, 229
Welsh plug, 229
Werkspoor, 162
Whirling, 12
Working point, 64, 87